THE EPOCHS OF NATURE

THE EPOCHS OF NATURE

GEORGES-LOUIS LECLERC,
LE COMTE DE BUFFON

Translated and Edited by
Jan Zalasiewicz, Anne-Sophie Milon,
and Mateusz Zalasiewicz

Introduction by
Jan Zalasiewicz, Sverker Sörlin, Libby Robin,
and Jacques Grinevald

Illustrations by Anne-Sophie Milon

THE UNIVERSITY OF CHICAGO PRESS CHICAGO AND LONDON

The University of Chicago Press, Chicago 60637
The University of Chicago Press, Ltd., London

Published 2018
Printed in the United States of America

27 26 25 24 23 22 21 20 19 18 1 2 3 4 5

ISBN-13: 978-0-226-39543-2 (cloth)
ISBN-13: 978-0-226-39557-9 (e-book)
DOI: https://doi.org/10.7208/chicago/9780226395579.001.0001

Library of Congress Cataloging-in-Publication Data

Names: Buffon, Georges Louis Leclerc, comte de, 1707–1788, author. | Zalasiewicz, J. A., translator. | Milon, Anne-Sophie, translator, illustrator. | Zalasiewicz, Mateusz, translator.
Title: The epochs of nature / Georges-Louis Leclerc, le comte de Buffon ; translated and edited by Jan Zalasiewicz, Anne-Sophie Milon, and Mateusz Zalasiewicz ; introduction by Jan Zalasiewicz, Sverker Sörlin, Libby Robin, and Jacques Grinevald ; illustrations by Anne-Sophie Milon.
Other titles: Époques de la nature. English
Description: Chicago ; London : The University of Chicago Press, 2018. | Includes bibliographical references and index.
Identifiers: LCCN 2017029443 | ISBN 9780226395432 (cloth : alk. paper) | ISBN 9780226395579 (e-book)
Subjects: LCSH: Natural history.
Classification: LCC QH45.B913 2018 | DDC 508—dc23
LC record available at https://lccn.loc.gov/2017029443

♾ This paper meets the requirements of ANSI/NISO Z39.48-1992 (Permanence of Paper).

Contents

Preface

This translation owes much to serendipity, although a consistent backdrop has been a transdisciplinary, and indeed international, exploration of the Anthropocene: the concept that we live in a new epoch where humans now drive many of the geological processes on Earth. By any practical contemporary measures, this concept was launched as an improvisation by the Nobel Prize–winning chemist Paul Crutzen at a meeting in Mexico in 2000. It has since grown rapidly in use and influence, both within Earth sciences and beyond. While the present context and its applications to understanding the future have been the emphasis of Anthropocene discussions, the idea of epochs has a long history. Even the idea of humanity as a driver of geological processes has a prequel: a first scientific argument came from none other than Buffon, who described the seventh and last epoch of his Earth history as reflecting "assistance by the power of man."

It was within the Anthropocene Working Group—the body currently examining the Anthropocene with regard to potential formalization within the Geological Time Scale—that one of the authors (JG) first made another (JZ) aware of Buffon's significance not just to this new field of study, but to the development of the science of Earth history in general. And it was as a result of Anthropocene conferences that links were made with other members of the team: at Sydney in 2014 (with LR and SS) and then in Toulouse, and subsequently at Nanterre, Paris (ASM). Serendipitous (and peripatetic) associations, certainly—but they became a reason to remedy a two-century-long oversight: that is, to provide the first complete translation into English of one of the shortest, yet most vivid, most wide-ranging and influential of all of Buffon's works.

A good deal of our motive was that this masterpiece by Buffon should, simply, be *more widely read*: not only by specialists within the history of science—though we hope this edition will be of use to that important community, too—but to scholars of both the sciences and humanities more generally.

It really is an extraordinary book—a very good read, even now. At the time, Buffon was criticized for making the best science of the day accessible to the general reader, but this is why it translates so well today. Buffon combines fine observational detail (and deductions therefrom) with an overarching vision of our planet's history, from beginning to end.

The translation was mainly done by JZ and ASM, with critical advice from JG—and from the reviewers and readers of this manuscript both formal and informal (including Philip Sloan, Noah Herringman, Adrian Rushton, and Colin Summerhayes), to whom we are deeply grateful. The many fragments of translated text were assembled by MZ into a coherent electronic document that could then be commented on and evolved by all the authors. The introduction was assembled by JZ, SS, LR, and JG and so retains viewpoints ranging from those of the professional geologist to historians of science and of the environment, and here there is diversity reflecting the specific significance that Buffon carries to each. The illustrations are by ASM. A particularly tricky piece of translation, of seventeenth-century dog Latin from Leibniz's *Protogaea* (in the "Notes justificatives"), was by done by Sara Jones and Adrian Rushton, whom we heartily thank—as we do Christie Henry, Kelly Finefrock-Creed, and their colleagues at the University of Chicago Press, who have been insightful and supportive throughout.

We have aimed at a translation that is faithful to the original in substance and style—though of course both cannot be perfectly preserved simultaneously. We try here to act as a bridge between the French language of the eighteenth century and an English language both modern and international. Buffon's famous style was aimed toward clear transmission of content, and not at constructing an elegant facade to hide or elaborate that content. His prose, hence, has a basic clarity that greatly helped translation, even across more than two centuries of the languages' evolution. But it was, of course, in the style of its day. Buffon's text contained, for instance, a great many more semicolons than make sense today, and so some liberties have been taken with the punctuation. Otherwise, we have tried to remain faithful to the original, including the use of archaic measures of Buffon's time, to which we provide a brief glossary below.

The hardest words to translate were of course the simplest, and we have decided to stay within the sensibilities of Buffon's world, with direct translations of words such as "sauvages" and "l'homme." This simple approach can elide nuance. Thus, for Buffon, "sauvage" was a being of the forest, not quite human and not quite animal, the word not carrying the overtly racist overtones of today. Another term, "moule interieur" is typically translated by historians of science as "internal mold," denoting one of Buffon's important concepts, of a means of guiding embryos in their development. To a paleontologist, though, the word "internal mold" is indelibly associated with a specific form of fossil preservation, while the most appropriate translation we hit upon, "internal matrix," was too far from the typical translation of the history of science community. We hence settled upon "inner mold" as a compromise translation.

We hope that such compromises do not devalue our end result too greatly, and that *le tout ensemble* gives an honest rendering of this extraordinary work.

JG, ASM, LR, SS, JZ, MZ

Translators' Note

The text translated here is the original version used by French historian of science Jacques Roger, Buffon's noted modern biographer, for his critical edition, published at Éditions du Muséum National d'Histoire Naturelle, Paris, in 1962. In addition to the text of *Les Époques de la Nature*, comprising an introductory "First Discourse" and a chapter for each of the seven epochs, we include a translation of Buffon's substantial "Notes justificatives," in which he provided, for certain of his statements in the text, details of what he considered important supporting evidence, in part from the correspondence and publications of others. These *Notes* include important and fascinating material on topics such as fossils and fossilization, the former presence (or not) of human giants on the Earth, and evidence of recent climate change and glacial extent, which throw additional light on Buffon's thinking. In this edition, the *Notes* are linked to the main text via superscript numerals, while we retain Buffon's few separate footnotes to the text as footnotes, designated by symbols. We confine our own footnotes to our introduction.

The *foot* referred to here is the French "king's foot" (royal foot) and is divided into 12 inches; they are about 7 percent longer than the feet and inches used today. The smaller unit "line" is one-twelfth of an inch.

1 *aune* is 3 royal feet.
1 *toise* is 6 royal feet, and equals 1.949 meters.
1 *fathom* is 1.83 meters.
1 *league* is 4.9 kilometers.
1 *quintal* is 48.95 kilograms.

The Cabinet du Roi was a property originally bought by Louis XIII in 1633, which became a cabinet of curiosities, then subsequently the Jardin du Roi, then the Jardin des Plantes and now the Muséum National d'Histoire Naturelle in Paris.

The "horn of Ammon" is the common fossil that we know today as the ammonite.

Other terminology: In our translation, with geographic names, we have tried to keep some balance between names in Buffon's time and those of today: hopefully most are clear from the context, though we have opted to retain "New Holland" instead of using the modern "Australia." There are a couple of specific terms in the text, too, that are worth noting here: they relate to scientific concepts—in themselves complex and variously debated—current in Buffon's time. One is the "inner mold," which relates to a kind of power within each type of organism that Buffon held to organize the "organic molecules" of that organism into their particular form. Another is "germs," which at the time were widely held to be tiny preformed individuals of the adult present within eggs and/or sperm. These, and related concepts, can of course be explored at length (Jacques Roger's biography of Buffon, noted below, is as good a place as any to start), though our main concern here is simply to present the narrative as Buffon told it.

Introduction: Buffon and the History of the Earth

JAN ZALASIEWICZ, SVERKER SÖRLIN,
LIBBY ROBIN, AND JACQUES GRINEVALD

The Significance of Buffon Today

Buffon, outside of the French-speaking world, remains somewhat in the shadows today. It is curious that until now there has been no complete English translation of *Les Époques*. Is that because Buffon became so quickly passé, a mere figure of the ancien régime, that he dropped out of sight with uncommon speed? Or is it, as his biographer Jacques Roger reflects, because he wrote too much, with those thirty-six volumes of the *Histoire naturelle*, mostly published in his lifetime? Or is it because he was too popular? He wrote French beautifully, evocatively, excitingly—and he was not afraid to take on big ideas. So, with more than a hint of sour grapes, some of his fellow savants spoke of him as a "phrasemonger," the inference being that such accessibility excluded him from the highest levels of the scientific elite.

Perhaps. For whatever reason, there is certainly not an excess of literature on him in the English-speaking world.* Buffon is sometimes prone to be cast, indeed, as a rather weighty member of a scientific establishment whose overgeneralizations act as springboard and foil for the dynamic new intellectual

* Jacques Roger, *Buffon: A Life in Natural History* (Ithaca, NY: Cornell University Press, 1997), is a major biography, while an essay in Stephen Gould, *The Lying Stones of Marrakech* (London: Jonathan Cape, 2000), is a vivid introduction. See also Otis Fellows and Stephen Milliken, *Buffon* (New York: Twayne Publishers, 1972); John Lyon and Phillip Sloan, eds., *From Natural History to the History of Nature: Readings from Buffon and His Critics* (Notre Dame, IN: Notre Dame University Press, 1981); J.-Cl. Beaune et al., eds., *Buffon 88: Actes du Colloque international pour le bicentenaire de la mort de Buffon (Paris-Montbard-Dijon, 14–22 June 1988)* (Paris: Librairie Philosophique VRIN, 1992); Kenneth Taylor, "Buffon, Desmarest, and the Ordering of Geological Events in *Époques*," in *The Age of the Earth: From 4004 BC to AD 2003*, ed. C. L. E. Lewis and S. J. Knell, Special Publication 190 (London: Geological Society, 2001), 39–49; Martin Rudwick, *Bursting the Limits of Time* (Chicago: University of Chicago Press, 2005); Martin Rudwick, *Earth's Deep History: How It Was Discovered and Why It Matters* (Chicago: University of Chicago Press, 2014). See also the website www.buffon.cnrs.f/.

generation, as were represented, for instance, in Keith Thomson's account of Thomas Jefferson as scientist.*

Why, then, bring Buffon's views on the Earth to an English-speaking audience, in full, now, almost a full quarter of a millennium after they originally appeared in print? One reason is that, in some ways, the sciences have come full circle. There has been more than two centuries of, for the most part, ever-increasing subdivision of this area of knowledge, and the development of disciplines, then subdisciplines, then specializations within subdisciplines. Now, it has become increasingly clear that one has to understand not just the parts (in minute detail) of the whole, but also the "whole" itself (an entity that we now know, even better than before, is not just the sum of its parts). Among the most important incarnations of the whole is that of this Earth, the planet that we still entirely rely upon for our existence. Buffon's vision of the Earth and—perhaps more particularly—the way he developed it, may have lessons for us yet.

Hence, some closer acquaintance with this savant of pre-Revolutionary France may be useful to a wide community, now that natural history and cosmology have reunited in planetary systems science, and in other developments such as Big History,† and where the many specialist sciences that have emerged in the intervening three centuries are now again working together. Not every polymath synthesizer of today can be assumed to be a reader of French, yet the ideas of Buffon take on a new significance as stratigraphers reconsider the idea of "epochs," consider anew how the present (and future) relate to the past—and ask the question whether the epoch of the Holocene is now completed, to be replaced by a yet newer epoch.

In 2008 the Stratigraphy Commission of the Geological Society of London considered this problem, in pondering the new term "Anthropocene" that had been launched into public debate only a few years previously by the Nobel Prize–winning atmospheric chemist Paul Crutzen within the Earth System science community that, perhaps more than any other, is trying to reweave the separate disciplines into a new understanding of our planet.‡ Now, the term is under scrutiny by geologists, indeed by *stratigraphers* (to revert to a disciplinary division—and indeed one that Buffon arguably helped to found), by a working group of the International Commission on Stratigraphy.

* Keith Thomson, *Jefferson's Shadow: The Story of His Science* (New Haven, CT: Yale University Press, 2012).

† David Christian, *Maps of Time: An Introduction to Big History* (Berkeley: University of California Press, 2004).

‡ P. J. Crutzen, "Geology of Mankind," *Nature* 415 (2002): 23; P. J. Crutzen and E. F. Stoermer, "The 'Anthropocene,'" *IGBP Newsletter* 41 (2000): 17–18.

Perhaps, though, it is a sign of the times that this working group is not made up exclusively of experts in stratigraphy, but includes also historians, geographers (sensu lato), archaeologists—and a lawyer.

There is some significance to the historical moment when *epochs* are first described. Buffon was writing just on the cusp of the French Revolution, and it was also the cusp of the Industrial Revolution. This—perhaps better represented as the *thermo-industrial* revolution, to more squarely place the accent on the unprecedented rise in (fossil) energy use*—is what Paul Crutzen fingered as the cause of anthropogenic planetary-scale environmental changes, which he (and Eugene Stoermer, a lake ecologist) called the Anthropocene. For many, the arcane deliberations of the International Commission on Stratigraphy are beside the point: the idea of the Anthropocene has moral, ethical, and personal implications for *all* the planet's citizens. Moving beyond stratigraphy (but including its essence), science journalist Elizabeth Kolbert writes of *The Sixth Extinction* in this Age of Humans.†

As the Working Group on the Anthropocene (part of the Subcommission on Quaternary Stratigraphy of the International Commission on Stratigraphy) deliberates, the concept has reverberated through the intellectual lives of artists and humanists, through biology and social sciences, through museums and other popular intellectual forums. The moral responses of people working at global and local scales—and every scale in between—have been reflected in events, performances, exhibitions, and justice protests. A concept that originated in the impulse "not to reduce the future to climate," in Mike Hulme's words, has undoubtedly expanded the possibilities for rethinking the future of life on Earth, and the role of humans in that future.‡

Buffon's legacy as a polymath, working through the evidence of his times to create a great narrative that shaped the future of many sciences (including geology) is important to all those engaged with the nature of global change. Indeed, specific links between the ancient worlds that Buffon explored and the rapidly changing world of the Anthropocene have been explored in detail

* Jacques Grinevald, "La révolution carnotienne: Thermodynamique, économie et idéologie," *Revue européenne des sciences sociales* (Cahiers Vilfredo Parieto), 36 (1976): 39–79; Grinevald, *La Biosphère de l'Anthropocène: Climat et pétrole, la double menace. Repères transdisciplinaires (1824–2007)* (Genève: Georg Editeur, coll. "Stratégies énergétiques, Biosphère et Societé," 2007).

† Elizabeth Kolbert, *The Sixth Extinction: An Unnatural History* (New York: Bloomsbury Publishing, 2014).

‡ Mike Hulme, "Reducing the Future to Climate: A Story of Climate Determinism and Reductionism," in *Klima*, ed. J. R. Fleming and Vladimir Jankovic, *Osiris* 26 (Chicago: University of Chicago Press), 245–66.

by Noah Heringman.* Debate today ranges from the planetary traces of the actions of humanity that stratigraphers identify in the rocks to patterns identified by social justice campaigners and activists in the unequal distribution of the bounties of the industrial revolution.

The meaning of concepts such as *eons*, *eras*, and *epochs* will undoubtedly continue to develop with future knowledge, but the long shadow of Buffon's original construction of the first of them, *Les Époques*, remains crucial to the scientific and moral arguments for and against a recognition of human traces in the strata of the Earth.

Would he have been surprised that his first Earth history would be discussed at the beginning of the third millennium—a time that would have seemed far distant from the standpoint of the Age of the Enlightenment? Perhaps not altogether, for he knew—and valued—his own worth. But his time was distant not only in simple, perceived temporal separation. Buffon's world—or rather Buffon's Earth—was one where the atmosphere still held a standard interglacial measure of 280 parts per million of carbon dioxide, rather than the 400 today, the latter being akin to that of the Pliocene Epoch, before humans walked the Earth. It still had nitrogen and phosphorus cycles that were undisturbed, rather than approximately doubled in scale. The biosphere, though long impacted by humans on land, was still almost pristine in the oceans—and its recent offshoot, the technosphere that both nurtures and binds us today, had barely begun its explosive development. As we read him now, to help gain context for our contemporary predicament, we must recall that we are reading words written on a different planet.

Georges-Louis Leclerc, le comte de Buffon

Georges-Louis Leclerc, le comte de Buffon. Two names, one man: a self-made man, by any standards. A self-made *aristocrat*, indeed, who climbed the social ladder the hard way, to become a power in the land and an influence well beyond it. He was multitalented, multitasking, and a naturalist. He had a long life—but, luckily for him, not too long. Born in 1707, he died in 1788, aged eighty, just before the French Revolution. Perhaps he took his optimistic notions of human progress to the grave. Had he lived a little longer, Madame Guillotine would likely have claimed him, as she took his son.†

* N. Heringman, "Deep Time at the Dawn of the Anthropocene," *Representations* 129 (2015): 56–85.

† Buffon's son, Buffonet—perhaps a revealing nickname—could not live up to the almost superhumanly high standards set by the father. His youth was short and somewhat erratic. He

Political fashion aside, Buffon was influential—Ernst Mayr called him the most important naturalist between Aristotle and Charles Darwin. He influenced James Hutton, Alexander von Humboldt, Charles Lyell, and Darwin, and also Vladimir Vernadsky and other scientists of later generations. He famously crossed swords with Thomas Jefferson, author of the Declaration of Independence and later the third president of the United States, over the status—majestic or enfeebled?—of American mammals, and over the identity of the bones of the mastodon. Their mutual respect was too deep for this to become a feud—they corresponded, and met amicably when Jefferson lived in Paris from 1784 as US ambassador.

Buffon had such influence partly by the extraordinary power, and coherence, and scale of his thinking. He set the questions for natural history, even where he was not in a position to answer them—and enjoined both his peers and the common people to consider them through his clear, vivid narrative. Buffon was one of the great stylists in any language. His fellow savants noted, at times somewhat disdainfully, that his writing could be read and absorbed not just by his peers, but by anyone. Not all his amateur readers were common: Catherine the Great of Russia was among them. He thus developed further the genre of what was later called "popular science," which had been pioneered by seventeenth-century savants like Galileo and Fontenelle, and by the middle of the eighteenth century had a growing number of practitioners elaborating on "Il Newtonianismo per le dame" (the title of Francesco Algarotti's famous book from 1737) and other topics for widening audiences.* Modern equivalents might include Stephen Jay Gould and Carl Sagan, as they, like Buffon, wrote about their own science, trying passionately to avoid compromises in conveying its reality. Natural history and natural philosophy in Buffon's day had not yet crystallized into the plethora of disciplines we know from the nineteenth and especially the twentieth centuries, each hypertrophied in their own materials and terminologies.† Even so, his narrative power was considerable. Lucid writing was as important to Buffon as philosophical logic: he could not do one without the other.

---

came of age, briefly, in the Revolution, and indeed joined the revolutionaries. This did not prevent him from being denounced and sent to the scaffold. His last words, facing the *tricoteuses*, were "Citizens, my name is Buffon."

* Jonathan I. Israel, *Radical Enlightenment: Philosophy and the Making of Modernity, 1650–1750* (Oxford: Oxford University Press, 2001); Patricia Phillips, *The Scientific Lady: A Social History of Woman's Scientific Interests, 1520–1918* (London: Weidenfeld & Nicolson, 1990).

† Nico Stehr and Peter Weingart, "Introduction," in *Practising Interdisciplinarity*, ed. Stehr and Weingart (Toronto: University of Toronto Press, 2000), xi–xvi.

Georges-Louis Leclerc was born, in 1707, into the kind of family whose successive members might figure in the slow and patient novels of a century ago. The Leclercs lived at Montbard, in the Bourgogne region of France. Generation by generation, through judicious choice of employment and spouse, they climbed gradually higher. Laborers first, most likely; then a barber-surgeon's apprentice, who earned enough to send his son to train as a doctor—and his son, in turn, became a local judge. This was Buffon's grandfather. Buffon's father became a lawyer—and bought the local rights to collect the unpopular (but lucrative) salt tax. Marriage improved the family prospects yet further—his wife's uncle was wealthy (also from tax collection). He was childless, too, and when the young Georges-Louis arrived into the world, it was a good move to ask him to be godfather. He died soon after, leaving his fortune to the infant: with it, the father bought for him the rights to the holding of Buffon, a small village a few miles away, and the lord's rights to the castle there—hence *comte de Buffon*.

The young Buffon, therefore, grew up into a tradition of careful and solid security: a lesson that he wasn't to forget. He could—he *should*—have grown up to become a person of some standing in local government. But, somehow, Buffon became one of the greatest naturalists of France. Jacques Roger, his modern biographer, would have him as *the* greatest, in the full knowledge of the redoubtable trio that made up the following generation—Jean-Baptiste Lamarck, Etienne Geoffroy Saint-Hilaire, and Georges Cuvier.*

Buffon trained in law like his father—but during his studies developed a taste for the ideas of natural philosophy then being discussed, avidly, among small groups of like-minded people. His father disapproved, not least because science was not a recognized profession, so even if he were highly successful, it would mean stepping down (or falling off) the social ladder. The young Georges-Louis proved him wrong, on both counts. Relations with his father seem never to have been good. They took a turn for the worse when the father, a widower, married a younger woman, jeopardizing George-Louis's inheritance from the rich uncle. George-Louis threatened his father with a lawsuit. It is not known whether the lawsuit went ahead, but the effect was, for the young Buffon, successful. He kept the Montbard estate, and its castle. This provided him with a base—and a workplace—for the rest of his life.

Buffon carefully cultivated his contacts and maintained a work ethic of twelve to fourteen hours a day sustained, day in, day out, for the rest of his life. He was not a natural workaholic: this was the effort of will of someone

* Roger, *Buffon: A Life*.

who admitted to sloth, a will enforced by hiring a laborer to drag him out of bed early in the morning.

He made his reputation first with mathematics, obtained a junior position at the Academy of Sciences, researched (at Montbard, largely) the properties of timber, then became head of the Jardin du Roi (the Royal Botanic Gardens) in Paris in 1739, at the age of twenty-two, a post he kept until his death, a remarkable directorship of almost precisely half a century, a term that ended only five years before the revolutionary government changed its name to the less royally associated Jardin des Plantes. He moved beyond mathematics and into the higher calling of natural history. He was also an effective administrator, both at the Jardin du Roi and at Montbard, which left him time to describe systematically everything, from minerals, plants, and animals, including humans, in his *Histoire naturelle*, amounting to thirty-six volumes in his lifetime and several posthumous ones. Voluminous in every sense, but brilliantly philosophical, it put him virtually on the same level as his prominent contemporaries Voltaire and Jean-Jacques Rousseau in public philosophical debate, although to posterity he has been seen more as a natural historian than as a philosopher. He focused mostly on the mysteries of *life*, asking questions that have continued to be discussed over the ensuing three centuries. For instance, he considered a species as a set of organisms that can interbreed and produce viable offspring, a basic principle still in use.*

His most widely read work, though, was shorter—a single slim volume—but the one that ranged most widely over time and space. *Les Époques de la Nature* (*The Story of the Earth*) was written late in his life. It was also the first full narrative of the Earth on natural history principles. It built on his earlier sketch, *La Theorie de la Terre*, published some four decades previously among the early volumes of the *Histoire naturelle*, as an early example of modern writing on geological successions. It was not quite the first stratigraphy, though. In preceding decades, the Italian Giovanni Arduino and the German Johann Gottlob Lehmann had seen that younger, softer, fossil-bearing rocks lay upon older, harder ones. Leonardo da Vinci had made a similar observation more than two centuries earlier. In *Les Époques*, Buffon ranged both more widely and more deeply than these and later studies, creating an original approach, exceptional in his time. The parallels of this text sometimes seem closer to, say, the rise of Earth System science in the late twentieth century, than to the patient nineteenth-century categorization of fossils and strata by

* Since introduced in its modern version by Ernst Mayr in 1942, although of course with many critiques and qualifications since then. Ernst Mayr, *The Growth of Biological Thought: Diversity, Evolution, and Inheritance* (Cambridge, MA: Belknap Press of Harvard University Press, 1982).

William Smith, Adam Sedgwick, Roderick Murchison, and Alcide D'Orbigny that gradually assembled, in piecemeal fashion, the Jurassic and Cambrian and Silurian and all the other giant rungs of the Geological Time Scale.*

In comparison with his peers, Buffon shared the same empirical evidence, but he was more philosophically ambitious with the data gleaned from fossils and minerals of the local strata. Buffon wanted to break out of the simple sequential Earth narrative and create nothing less than a fully working and integrated model of the Earth, from beginning, through present, to the end, where the Earth is categorically shown as one planet among other planets. In such a vision (and much of it was visionary, given the enormous gaps in the evidence base) everything must fit together with, at least, self-contained logic, and the story must be consistent with all the available facts. The interaction of planetary rock, atmosphere, ocean, topography, sea level, volcanoes—and life itself—must interact in a way that made *sense*.

Geologists in our time are used to creating their narrative, in a forensic manner, from highly fragmentary evidence. A typical geological map, for instance, shows a comprehensive model of the rock formations, complete with predictions of underground structure, from scattered rock exposures that can often represent less than one percent of the area represented. Such a map is emphatically an interpretation (a four-dimensional one, for it perforce also includes interpretation of the geological evolution of the chosen area, all compressed onto the two dimensions of a sheet of paper). The intention of such a map is not to disclose eternal verities, but to a show a working model, one that is subject to change as more evidence is unearthed. Buffon was operating in this fashion, not only with far scantier evidence at hand, but also without much in the way of tested, preexisting conceptual frameworks. Indeed, he built a good deal of the conceptual framework in order to write his narrative.

Hence, he was bound to get lots of things wrong. For instance, the material of the Earth was not torn out of the Sun by a comet (an idea that Buffon had inherited from Newton,† just as Newton had also entertained a cooling theory

* The history of how the early savants wrestled with the scale of geological time and process has been marvelously described in Martin J. S. Rudwick's, *Bursting the Limits of Time* and *Worlds before Adam* (Chicago: University of Chicago Press, 2008). Those who have not time to read through the combined 1,322 pages of these definitive works might try the elegant summary in Rudwick's *Earth's Deep History*.

† Isaac Newton brought this up in a letter to Thomas Burnet written in 1680 or 1681, in quite amusing language: "Now for the number & length of the six days: by what is said above you may make the first day as long as you please, & the second day too if there was no diurnal motion till there was a terraqueous globe, that is till towards the end of that days work. And then if you will

and the idea that the days of Creation were not actually "days" but could span very long time periods); the Sun does not shine because of the tidal effects generated by the comets and planets that spin around it; mountain ranges are not the crumples left over from the initial cooling of a solidifying Earth; sea levels do not fall by some kind of tumbling of oceans into enormous collapsing caverns (an idea Buffon may have received from Descartes in *Principia Philosophiae*, 1644, or from Thomas Burnet in his *Telluris Theoria Sacra*); volcanoes are not powered by exothermic (heat-producing) reactions between inrushing seawater and subterranean minerals. These are, as we now perceive the history of the Earth, all first-order misinterpretations that Buffon made of individual planetary phenomena.

But Buffon was attempting to make coherent sense of a world that had only just begun to be explored with the methods of modern natural history and in terms of a set of emerging young specialized sciences. These were days before the dawn of most of the fundamental concepts and models that we now use to organize scientific knowledge. The chemical elements Buffon spoke of, for instance, were earth and fire and water—not carbon, oxygen, and silicon, and he had neither an atomic theory nor a periodic table at his disposal. This did not stop him thinking through the scientific consequences of what he could see. He was an avid reader and an effective correspondent, so his ideas included much of what others in his class saw across the rest of the known world. And it did not stop him calculating some of the forces and time scales involved, using the reasonably well-developed mathematics and astronomical observations of his day. From this perspective, some of his speculations may seem less preposterous.

In Buffon's interpretation of our planet's overall history, though, some features were fundamental, and provided the framework upon which everything else was built. The Earth, thus, had a beginning that could be explained using physical principles, then went through a long succession of changes before humans appeared, and is changing still—partly under the influence of humans, he emphasized—and will one day come to an end. He thus stood in opposition to the conservative orthodoxy proposed by Linnaeus, who argued a divine unchangeable, steady state of the living world, with a fixed number

suppose the earth put in motion by an eaven force applied to it, & that the first revolution was done in one of our years, in the time of another year there would be three revolutions of a third five of a fourth seaven &c & of the 183[d] yeare 365 revolutions, that is as many as there are days in our year & in all this time Adams life would be increased but about 90 of our years, which is no such great business" (http://www.newtonproject.ox.ac.uk/catalogue/record/THEM00253). Original document in Keynes Ms. 106(B), King's College, Cambridge, UK.

of unchangeable species, and his contemporary geologists who argued only a slow and gradualist change of the Earth itself.* But he also stood as the latest in a line of predecessors who had initiated speculation about a dynamic, historical Earth, where major events had occurred with or without a framing biblical chronology, from Descartes to James Ussher, Athanasius Kircher (*Mundus subterraneus*, 1655–78), Thomas Burnet (*A Sacred Theory of the Earth*, 1681–89), Johann Scheuchzer (*Herbarium diluvianum*, 1709, and *Physica sacra*, 1731–35), and James Hutton, who in earnest opened the possibility of a geological deep time.†

To Buffon, critically, the Earth was not "timeless": he suggested, indeed, that the overall shape of continents, oceans, and mountain ranges was in part preserved from its primordial state on cooling, which according to his own experimentation at Montbard he proposed had been going on for close to seventy-five thousand years. This evocative vision was made immediate by his narrative and descriptive skill.

Indeed, his vision of Earth was a little like ours is now of Mars, where we can still see, in the current topography of that planet, traces of its most ancient history—for instance via the hypothesis that the basic distinction between the Martian southern highlands and northern lowlands might reflect a very large asteroid impact very early in that planet's history. So while Buffon's vision is far from the image of Earth history that later developed, it had surprisingly many of the general features, and above all it was a coherent statement, log-

* Philip R. Sloan, "The Buffon-Linnaeus Controversy," *Isis* 67 (September 1976): 356–75; John C. Greene, *Science, Ideology, and World View: Essays in the History of Evolutionary Ideas* (Los Angeles: University of California Press, 1981), 37–40. See also Lyon and Sloan, *From Natural History to the History of Nature*. Mary Efrosini Gregory, in *Diderot and the Metamorphosis of Species* (London: Routledge, 2006), shows how Buffon's heterodox ideas not only clashed with those of Linnaeus but also how they were taken even further by his contemporary Diderot. Buffon, although seemingly modern in the early twenty-first century, was much criticized by some of his French contemporaries, for example, by Malesherbes and Voltaire, and by d'Alembert who called him "a great phrasemaker," whereas in fact the criticisms from the Church were fairly modest; they have become aggrandized in later eulogisms of Buffon. It is necessary to keep in mind when it comes to the "forerunner" status often attributed to Buffon, not only in relation to the modern concept of species but also as an early proponent of evolutionism, that these uses of Buffon have in themselves a history that has served a purpose in later political and scientific controversies, for example, on the role of French science pre-Darwinist evolutionary theory. See, for example, Stuart Persell, "The Revival of Buffon in the Early Third Republic," *Biography* 14, 1 (1991): 12–24. A modern critique of precursor narratives is Clive Hamilton and Jacques Grinevald, "Was the Anthropocene Anticipated?," *The Anthropocene Review* 2, 1 (2015): 1–14.

† Martin J. S. Rudwick, *Scenes from Deep Time: Early Pictorial Representations of the Prehistoric World* (Chicago: University of Chicago Press, 1992); Rudwick, *Earth's Deep History*.

ical in itself. It did not have the benefit of the successive twentieth-century understandings of continental drift and plate tectonics that are now central to geology. Nevertheless, his Earth was ancient, and dynamic, and its workings and history were constrained by the evidence of its component materials, both inanimate and living.

The Earth he portrays was, for its day, radically, and in some quarters blasphemously, old. For Buffon personally, it was *dangerously* old, given the power still held by the Sorbonne, then a conservative theological college quite unlike the institution it was later to evolve into. This might help explain the structure of *Les Époques*. There is a "First Discourse," a kind of introduction, then the main text—the seven epochs given a chapter each—followed by a considerable number of detailed *notes justificatives* (justifying notes)—in which evidence to support the narrative is provided, chapter by chapter (Buffon included a useful glossary for his readers, too).

Part of the reason to have the First Discourse was to get the apologies in first. Buffon was writing when religious orthodoxy held considerable sway, and when it was dangerous to one's career—even for one as well-connected and politically astute as Buffon—to disseminate ideas that ran counter to prevailing biblical interpretation. Some argued that Buffon scorned the religious hierarchy and wrote his apologia in carefree irony—and parts of this section can be interpreted as provocatively written. But his biographer, Jacques Roger, suggested he was genuinely concerned to placate the powerful Sorbonne, then the main theological college in France as well as a significant place for Enlightenment scholarship.* Thus, after working hard to argue that the biblical time scale—that he was about to shatter—was written *metaphorically* rather than literally (i.e., that what in Genesis is called "days" of creation should in reality, for God, be seen as much longer time periods and, each of them, uneven in duration), Buffon wrote that his "purely hypothetical" ideas concerning the Earth could in no way harm the "unchanging axioms" of religious faith, which were "independent of all hypothesis." The stratagem worked, on the whole—though it did not stop criticism from some in the Sorbonne, which perhaps gave Buffon some anxious moments. Nevertheless, with these statements of overt piety out of the way, he then simply moved on to produce an empirically founded Earth history.

The age of the Earth was dealt with in two ways, which foreshadowed the numerical (or "absolute") and relative time scales of classical geological stratigraphy. His relative time scale was integrated with the numerical one that he devised, but was largely independent of it. Like the stratigraphies that fol-

* Roger, *Buffon: A Life*, 405.

lowed it, it was based on the evidence in the rocks, which he interpreted, quite unequivocally, to say that there had been an ancient Earth before people, and that it had different patterns of land and sea, different climate, and different biology—and that the ancient plants and animals, once gone, had disappeared forever. Thus he explicitly stated the notion of extinction, a generation before Georges Cuvier, the man commonly credited with establishing this idea, drawing heavily not only on Buffon's inspiration (chiefly from *Histoire naturelle*) but also on German comparative anatomists such as Friedrich Kielmayer and, in turn, his teacher Johann Friedrich Blumenbach.*

Already in his First Discourse, Buffon noted that in rock strata there were the remains of animals and plants that could not be found in nearby land or in adjoining seas. Therefore, these had either died out or moved elsewhere on Earth. So there, at the beginning of the main text, there is a modest suggestion, hedged about with some caution. In the "Notes justificatives," though, the equivocation disappears. He quotes the "large petrified volutes" and "horns of Ammon" (i.e., ammonites), that were "up to several feet across," "bélemnites," "numismales" (nummulites), and other such that were common in the limestones around Paris. The significance of these, he noted, depended on "long study and reflective comparison of all of the species of petrifactions found in the heart of the Earth": that is, he was looking forward to the start of a science, not yet born, that came to be paleontology. *Nevertheless*, "these examples, and others I can cite, are sufficient to prove that species of shells and crustaceans used to be present in the sea that do not exist any longer." You can't have a clearer—and more reasonably founded—statement than that. Robert Hooke had come to a similar conclusion in the late seventeenth century, regarding fossil ammonites, but Buffon developed the idea much further and it became a central idea underpinning his entire narrative.

He also detailed at length reports of enormous fossil skeletons pulled from the swamps adjoining the Ohio river in North America. These had bones and tusks (of "very good ivory") resembling those of an elephant—although the teeth were quite different, without complex grinding surfaces, but terminating in five or six blunt points, thus being more like those of a hippopotamus. After considering, then rejecting, the notion that these might represent a mixture of elephant bones and hippopotamus teeth (among many bones, none like the hippopotamus were found), Buffon concluded that this was an animal that had not survived to the present. For "an animal that is larger than an elephant cannot hide anywhere on Earth and still remain unknown," a position that

* Mark Barrow, *Nature's Ghosts: Confronting Extinction from the Age of Jefferson to the Age of Ecology* (Chicago: University of Chicago Press, 2009), 38–42.

provoked the patriotic sentiments of Thomas Jefferson, who kept believing in his cherished *incognitum*, proposed to be six times the size of an elephant.* This was pretty much the argument that Cuvier later, and influentially, applied to the mammoth. Here was Buffon using this logic two decades (and one political revolution) earlier, on what we now know as the mastodon. He also used the extinction of a species that had never been observed in order to sustain, ingeniously, the argument of his own theory of the Earth.†

There's more to *Les Époques*, though, than a flash of paleontological insight (a lucky hit, some might say). In this first attempted secular history of the Earth (and of the planets, indeed), from beginning to end, time's arrow flew inexorably from the white heat of (non-divine) creation to envisage a future Earth, frozen and biologically dead. It is a history derived from the evidence of the ground, some seen by him personally, and the rest taken from his prodigious reading and correspondence. A good deal of the narrative was not original to him, but his was an original synthesis that wove together ideas in circulation by word of mouth or in print. *Le tout ensemble* is his alone. The evidence demonstrated clearly that Bishop Ussher's much-debated few thousand years did not come close to being sufficient. The Earth had to be older. The question was how much older?

Buffon's measuring stick was essentially the same as that later used by Lord Kelvin—the cooling of the Earth. In his early trial run at *Les Époques*, *La Theorie de la Terre*, one of the first volumes of his *Histoire naturelle*, he had, in effect, an Earth without a history, without a beginning or end (somewhat akin to the vision his approximate contemporary James Hutton developed in his *Theory of the Earth*, published in 1788, which famously concluded that "we find no vestige of a beginning, no prospect of an end"). Within a rather vague time scale, land and sea, now and then, changed places. Later, though, Buffon was persuaded (through reading the arguments of Gottfried Wilhelm Leibniz) of an originally molten state of the Earth. That fitted in with known evidence (such as that temperature increased upon descending into mines underground) and with the overall vision that he was developing. Moreover, it gave him a measuring stick for Earth time.

How to calibrate that measuring stick, though? He heated up variously sized balls of iron and measured how long they took to cool. Measuring the temperature of such objects was not so simple then. Buffon did not trust the

* Barrow, *Nature's Ghosts*, 17.

† Stephen M. Rowland, "Thomas Jefferson, Extinction and the Evolving View of Earth History," in *The Revolution in Geology from the Renaissance to the Enlightenment*, ed. Gary D. Rosenberg (Boulder, CO: Geological Society of America, 2010), 227.

crude thermometers of the day—when theories of energetics were not well developed and phlogiston ideas were still current—so he measured instead the time it took for the ball to cool sufficiently to be held by hand for a minute without injury. (Buffon regarded women's hands, being more sensitive, as the most precise measuring devices.)

Projecting his data gave him a figure of seventy-five thousand years since the Earth had formed as a molten globe. He was aware that the margins for error were very large, and so he was very deliberately conservative. But, even so, that gave a starting point for the whole narrative. He, and the Earth itself, could now begin.

How to construct an Earth? In Buffon's chosen process, the first epoch begins with a comet striking a glancing blow against the Sun, the material flung out then condensing as the planets, which therefore start their existence as molten masses surrounded by vapor. They subsequently cooled—but how did the Sun stay white-hot? Buffon did not consider it as, for instance, a stupendously large burning coal ball. For him, it was the effect of all the bodies of the solar system, seen and unseen, orbiting around it. He thus seems to be invoking gravitational stretching and squashing. Buffon was wrong (the effect on the Sun is trivial)—but he was thoughtfully and interestingly wrong: we now know this is the mechanism that provides the heat energy to keep, for example, the spectacular volcanoes of Io, the moon of Jupiter, constantly erupting.

In the second epoch, the Earth cools, and begins to solidify. A crust develops on the surface, and this develops wrinkles and ridges—these are the present mountain chains—and, beneath those, bubbles that become underground caverns and cavities (he was to need those, later). To us now, it may seem that he had actually gone backwards from a steady-state Huttonian Earth where mountain ranges rose and fell, to a single-cycle Earth that has retained its primordial contours to the present day.

That would be a little unfair. In creating the first science-based whole-Earth narrative, he was positing a logically consistent succession of different states, through his empirically determined time scale. This fixed time scale could allow only a single basic geography, and so this is far from Hutton's "deep time." Nevertheless, to a child beginning to swim, the shallow end of a pool may seem uncomfortably deep. Even with his seventy-five thousand years, Buffon went out of his way to reassure his readers (who measured time in hours and years, and, perhaps, were aware of biblical time scales), suggesting ways to them of mentally coping with the unimaginable temporal abyss of those seventy-five millennia (think in terms of money, he said, and not years).

It took, he said—extrapolating from his iron-ball experiments—2,936 years for the Earth to solidify. Even as he wrote, he knew these figures were likely off by orders of magnitude: his unpublished manuscripts show that he was stretching his overall time scale some fortyfold, to three million years. But, he maintained his precise conservatism: partly, perhaps, to avoid his readers contemplating yet more outlandish time spans, partly to be prudent vis-à-vis the watchful Sorbonne, and partly to keep the narrative speeding along.

For the story does move on, with considerable élan. In its original version, shorn of the notes, glossary—and without Roger's illuminating but lengthy commentary in the widely recommended 1962 edition—it was just a shade over two hundred pages long—and that is with large print and small pages.*

Buffon evoked a picture of a solid, but still hot, Earth wreathed in water vapor, a jagged, barren landscape formed of igneous ("vitrescible") rocks. Below, precious metal ores form within rock fractures. Many details are given of this aspect; rather more, indeed, than on the vitrescible rocks themselves, for metals were big business, then as now, and a good early school for practical geology.

Some thirty to thirty-five thousand years after the Earth formed, he reckoned, it was cool enough for the gathering rainfall to begin to settle on the surface, without instantly being vaporized, as thick mists swirled, and tempests raged. And what an ocean formed! Buffon took the information he had—that strata with fossils could be found on mountains up to four thousand meters high—and that is where he placed the primordial sea level.

During his third epoch, the Earth was a waterworld—though one where the broad-brush geology is perfectly sensible. There was no debate here between plutonism and neptunism. Buffon simply stated, matter-of-factly, that the primary rocks are broken down, decomposed by the water to produce the salts in the world ocean and the sands and muds that accumulate in layers on the sea floor. Furthermore, he saw the link between muds and shales and slates, attributing the various states of these strata to different degrees of drying and compaction (his one-way history does not, though, allow for much in the way of metamorphic process).

Buffon applied what is today called lithostratigraphy—the discipline of arranging stratal units by their physical character—based on that of the hills and valleys around his beloved Montbard. He was clear that, in that region, there are layers of shale, overlain geometrically and succeeded in time by limestone.

* Jacques Roger, ed., *Buffon: Les Époques de la Nature*, Mémoires du Muséum National d'Histoire Naturelle, Nouvelle Série, Série C, Sciences de la Terre, Tome X (Paris: Editions du Muséum, 1962).

He describes, for instance, the three-dimensional connection between a well sunk in a valley (through fifty feet of shale) and the layers of limestone in the valley sides above. The limestone is full of shells, so the stuff of the rock, therefore, is made of the remains of countless generations of ancient animals; they extracted their shell-material out of the waters in which they lived (into which it had previously been put by the action of the waters upon the primordial fire rock). From a savant of the ancien régime, this appears as very modern-seeming sedimentary geochemistry.

He saw that the shales contain many fossils, too—those ammonites and belemnites. Life, therefore, appeared in his history pretty much together with the formation of sedimentary strata. The organic particles of which life is made, he thought, more or less automatically formed themselves into complex organisms, as soon as conditions became tolerable for life. There is no long gestation period for organic molecules here, still less any notion that it is the smallest, simplest kind of life that comes first. Life is thus an inevitable, and immediate, outcome of chemistry—on the Earth and, he said, on other planets too.

Despite being mentioned by Darwin and, as noted above, sometimes aggressively used later by scientists to claim French ownership in the prehistory of evolution, Buffon was not an evolutionist in the modern sense or, indeed, really at all. He inserted a dynamic, long-term time dimension into Earth history and he relativized the static species concept, two elements that were needed for later selection and evolutionary theory to occur, but that does not make him an evolutionist, and, hence, the Church could not accuse him of being one! He was, however, a biostratigrapher, for he could see that, in any one place, there was a succession of strata and fossils (the "elephant bones" from the surface sediments he knew came later than the ammonites). Perhaps more exactly, he was a biogeographer, charting the course of life as it followed the conditions of a changing Earth.

The pattern in such a model was—*must be*—clear and logical. The first to cool were the polar regions, and this is where rain first falls, and the oceans first gather. The cooling proceeded toward the equator, and the watery and habitable zone followed. Indeed, in one of Buffon's more spectacular imagined processes, he saw the southern tips of South America and southern Africa as having been carved by the ocean waters, as they rushed northward from the southern ocean (he had no idea, then, that those waters might conceal an Antarctica). Buffon's musings on the potential geomorphological power of such primary downpours might, in a sense, have been echoed by planetary scientists much more recently, as they worked to determine whether the primordial landscape of Mars (the outlines of which can still be glimpsed, more than three billion years on) was essentially shaped by a billion years of a more or less constant hydrological cycle of a warm, wet early planet, or by brief, brutal flood events interrupting what has almost always been a dry, frozen planetary surface.

For Buffon, the Earth's polar regions saw complex and abundant life while the low latitudes remained fiercely hot and inhospitable, an idea particularly appreciated by Catherine the Great, who admired not only Diderot and Voltaire but also Buffon. As the Earth cooled, the polar regions gradually congealed, and the baton of life was passed on toward the equatorial tropics. Species of animals and planets migrated, became extinct, or came into existence—assembled ready formed from organic particles—as the Earth's climate belts migrated. Somewhere in there, perhaps, lie the beginnings of paleoclimatology.

That might be stretching the case a little. But there is, in Buffon's account of his third epoch, some astonishing paleoenvironmental reconstruction. For he was aware that, as well as the ammonite-bearing shales laid down in the ancient ocean, elsewhere there were strata containing coal seams. He was aware that the coal strata and the marine shales tended to be tilted (which he thought was due to the sedimentary layers accumulating on steep slopes) and near-horizontal, respectively, and he guessed correctly that that the marine shales lay on top. He could see, too, that the coal-bearing strata contained many impressions of plants that looked "tropical" in nature, hence fitting in nicely with his cooling trend.

The coals, he went on, were the remains of the Earth's first vegetation, swept from the mountaintops that poked above the water, and into the sediment layers that surrounded them. Successive layers of plant debris and mineral sediment accumulated to form the many layers of coal in these beds. He wondered at the immense amount of plant material that grew and was buried, and mused on the immensity of past time that it must represent. More: he expressly compared these ancient coal-accumulating environments with the mouths of the Mississippi and the Amazon—and then (in some detail) with the coastal swamps of Guyana, where trees live and die and fall into the morass, there to decay. They fell so frequently in those thick forests, he noted, that travelers needed to be careful to sleep next to healthy and not rotting trees, to avoid being crushed in their sleep. As an evocation of what we would now call a modern analogue, it is highly effective, even today, and all the more remarkable as he did not see either the Mississippi or Guyana personally. He traveled there simply in spirit, through his extensive correspondence and his voluminous reading. The mind, he said, is the best crucible.

The crucible, though, works on what it can see. As Buffon ascribed the tilting of the lower sedimentary layers to their accumulating on steep slopes, it would follow that as the steeper slopes were gradually clothed in the sedimentary layers, the ancient topography would be smoothed, and so the newer strata would slowly tend to become more horizontal. With only scattered glimpses of bare rock in the vegetated and soil-covered French landscape, that was an allowable—even the most reasonable—hypothesis. A few years later, James Hutton, exploring the rocks of Scotland, was to find upended layers of strata planed off by erosion and then covered by younger strata: it was the sight of this structure, an angular unconformity, that gave him the concept of successive cycles of the building and erosion of mountain ranges and hence a temporal scale so enormous that it dwarfed Buffon's—making, as Hutton's friend and amanuensis, John Playfair, put it: "the mind . . . grow giddy by looking so far into the abyss of time."*

Such insights need the luck of seeing exactly the right rock exposure, and also context (Hutton was independently puzzling on how the Earth might work as an engine of heat and work, and so was attuned to making sense of his discovery). Buffon was working in different terrain, to a different "one-cycle" ground plan, and arguably had stretched time quite enough as it was for the society of his time and place to tolerate. Had he, somewhere in his local travels, encountered an unconformity such as the one that Hutton saw, one might imagine him neatly accommodating it into his overall one-cycle narrative (through a local effect of the crustal foundering he envisaged, for instance).

In his fourth epoch, the waters receded, and the landmasses, draped with fossil-bearing strata, were exposed. As the roofs of caves and caverns—those bubbles in the cooling crust—cracked and foundered, accompanied by earthquakes, allowing the waters to drain downward, Buffon inferred that the water went underground. (Later planetary scientists have entertained similar ideas concerning the fate of Mars's original water.)

When the sea level dropped, another phenomenon began: volcanism. This is not just Buffon trying to please the book-buying public, to pack as much of the Earth's genuine melodrama as possible into his narrative. It was his deduction of cause and effect in Earth processes, based on imperfect information, mostly drawn from secondary sources. He knew that many active volcanoes were at or near sea level—Stromboli, Etna, and so on. He also knew that there were extinct volcanoes in France, inland on high ground in the Auvergne. This was before biogeographer-explorer Alexander von Humboldt had ex-

* John Playfair, "Biographical Account of the Late Dr. James Hutton, FRS, Edinburgh," *Proceedings of the Royal Society of Edinburgh*, vol. 5, part 3 (1805 [read 1803]): 39–99, quotation on 73.

plored the high Andes, and brought back his famous reports of Chimborazo and Cotopaxi.

Therefore, Buffon reasoned, the mechanism that produces volcanic eruptions has something to do with the proximity of rock, air, and water. He surmised the spontaneous, catastrophic combustion of minerals such as pyrite. This detail was wrong, of course, but a logical explanation of the available facts. The most violent eruptions are often of the kind that volcanologists now term phreatomagmatic, whereby the flash-heating of seawater, suddenly introduced into a magma chamber, greatly increases the explosive force of an eruption. The hypothesis, in any event, gave his pen free rein to indulge in colorful descriptions of a ravaged Earth with emerging (and foundering) landscapes, vanishing seas, and volcanoes. Once the general mayhem died down, though, the world, now said to be just fifteen thousand years ago, emerged into the fifth epoch, with new lands: the northern kingdom of the giant elephants.

Buffon knew of the many finds of bones that resembled elephant, rhinoceros, hippopotamus, and such in Europe—and of the stories of similar bones, extracted in large numbers from the frozen lands of Siberia. He more or less ignored the common assumption that their remains had been swept into those regions by Noah's Flood. He knew there were simply too many of these skeletons for such an explanation to be true. And—other than his carefully crafted assertions, aimed at the Sorbonne, of the absolute primacy of divine scripture—he did not try to look for evidence of biblical events in the strata—and criticized those who did. Jacques Roger discussed at length the evidence of what religious belief Buffon may or may not have had. He wrote smartly and tactically to please the believers and the Church authorities, but on a deeper level Buffon did not give much away in his writings, and he attended Mass regularly, because that was the thing to do. Roger's view is that privately he was by and large an agnostic. Buffon simply interpreted the evidence in terms of physical, chemical, and biological processes.

For him, this was evidence of a warmer Earth—still unbearably hot in the tropics, he thought, but with a tropical-style fauna inhabiting northern lands that are now mostly frozen wastes. And those lands might have been even hotter, because some of the Siberian bones were larger than those of modern elephants. Thus, although he did not distinguish elephant from mammoth (as Cuvier was to do later) he did note differences. He ascribed these to what we might today call ecophenotypic variation, with morphology controlled by environment (in this case, temperature). This fitted logically within his overarching narrative of a cooling Earth.

Among these giant bones of this epoch were those of Buffon's "mystery animal," later identified as the mastodon. This, and an earlier assertion of Buffon's that the animals of America were relatively puny, slothful, and enfeebled by comparison with the rest of the world (a logical consequence, as he saw it, of the Americas being colonized relatively late in his Earth history), had acted as call to arms for Thomas Jefferson, who included the natural history of his native Virginia among his many interests. Stung to defend the reputation of North American mammals, Jefferson gathered, and sent to Buffon, a mass of information confirming their general strength and robustness—and also sent him the bones and skin of a large American moose (which arrived in Paris in poor—and smelly—condition) to underline the point. Jefferson was fascinated by the "Ohio bones," and saw more clearly than did Buffon that the teeth and the tusks belonged to the same animal. As a profoundly religious man, though, Jefferson did not believe that Providence could cast aside one of its creations, and so imagined the giant creature still present in some unexplored corner of the new continent; on this aspect, Buffon's intuitions were proved closer to the mark. Buffon seems to have accepted Jefferson's correction on North American mammals equably; the two men evidently respected each other.

With the sixth epoch there came the separation of the continents. For the bones of the "elephants" were scattered across Europe, Asia, and North America. Thus, he deduced there must have been free movement between these continents. It is not continental drift that Buffon is invoking here. Rather, it is ocean formation—that of the Atlantic, in particular—with Buffon proposing former connections between North America, Greenland, Scotland, and Scandinavia. It is another episode of crustal foundering that he saw as the cause. Islands such as the Azores and Newfoundland were seen as remnants of a former landmass, the great 1755 Lisbon earthquake is mentioned as, in effect, an aftershock of these larger crustal displacements, and the legend of Atlantis is also invoked.

The Earth's new geography, though, remained old in human terms. For he could see the new, later stratigraphy was building: the Nile and Mississippi deltas, and the coastal plain of Guyana, built of river muds. These new (and enormous) masses of sediment must, he saw, have postdated the birth of the Atlantic. As we get nearer to the present, Buffon's time scale is not so far off modern late Pleistocene-Holocene chronology.

These new landscapes were the foundation of the seventh epoch: the epoch of mankind. It has a curious title: *Lorsque la puissance de l'homme a secondé celle de la Nature.* That does not mean that the power of human actions on the Earth were secondary to natural forces—but rather that they assisted them. Here, humans arise as an animal species set apart from all others, and begin to transform the world. It is the first real expression of an idea that bears much resemblance to what in the twenty-first century is referred to via the Anthropocene concept. It is mixed in with history that is both real (Egyptian, Chinese) and very speculative, the existence of a much earlier, peaceful, and enlightened civilization, although not unlike speculations among utopianists, from Thomas More onward and certainly common in the vast literature in Buffon's own time on the *bons sauvages*. Generally Buffon saw human influence on the Earth as a good thing—not only in itself, but because humans could, for a while, warm the world and stave off the final, terminal freeze. But it is an influence that comes with a sincere responsibility. Since the strongest sentiment was fear, combined with the propensity for entertainment, human ingenuity had been directed to military violence and superficial divertissements. Humanity could do so much more. Also, the habit of going across the ocean and colonizing other peoples he saw as a bad one. Each people had the same right and capacity to develop and find their way to happiness. He finishes as an optimist, looking forward to humanity seeking glory not through war or colonial expansionism but through Enlightenment thinking. With power should come responsibility.

Les Époques is an epic, in scope if not in length. There was something to please—and to annoy—everybody. The reception it received was mixed—and, among Buffon's fellow savants, generally critical. The atmosphere is nicely given by a letter written (and perhaps sent) by Jean Etienne Guettard, a botanist and mineralogist (Buffon quoted him in *Les Époques*) who first recognized the Auvergne volcanoes as such and, in plotting the locations of minerals and their enclosing rocks in France, made a geological map, in 1746—more than half a century before William Smith's celebrated geological map of Britain (though he lacked Smith's stratigraphical insights). Guettard could have been referring only to *Les Époques* when he wrote: "Yet more Buffonades, my dear Count," it begins, going on to first faintly praise the "delicate and elegant phrases" with which Buffon, "with brilliant spirit, like Syrano [*sic*] de Bergerac," traced his "hypothetical ideas." Guettard then made clear that he thought this brilliance was put to ends that were not worthy of "the great Buffon" who was now "incorrigible, and that is not good, my dear Count, that is not good." The fine adventure story, he went on, warming to his theme, "would be devoured by the maidservant and then amuse the lackey—but it was unworthy of one who could shine a light for the most sublime Spirits."* In truth, Buffon's optimistic cosmology of his first epoch offered an easy target. Even as he was writing, realization was dawning that comets were not objects that were sufficiently large and dense to tear planet-sized masses of

* The letter is quoted in full in Roger, *Buffon: Les Époques*, cxxxix.

material from the Sun. And, there were complaints, as predicted, from among the theologians, who realized the significance of Buffon's time scale for the literal interpretation of the scriptures. But, these were an annoyance rather than a danger for Buffon—who, in any event, was hard at work on continuing the *Histoire naturelle*, with the world of minerals his latest task.

And, *Les Époques* did bring him a new readership that went far beyond the servant classes that Guettard spurned. Catherine the Great of Russia ordered a copy, and was captivated—not least because her Siberia was shown as a cradle of life on land—and corresponded with Buffon over the next two years. Buffon was delighted and replied in glowing (if not downright flattering) tone to the formidable monarch.

As royalty went into sharp decline just after Buffon's death in 1788, so did his reputation, at least in France. Even in natural history, these things matter, and the reputation of the man ennobled by Louis XV suffered with respect to that rising man of the people, Citizen Linnaeus, who had died a decade earlier and in whose honor societies were founded and monuments erected in an era obsessed with natural history.[*] And, of course, there came the wave of the new with Cuvier et al., to eclipse the works of the past.[†]

Buffon might largely rest, today, in the shade of his illustrious successors. But he deserves to be remembered, not least for his assembly of a truly holistic account of the Earth and its inhabitants. He did this, too, in style. Indeed, the phrase he is probably most associated with now is "le style c'est l'homme même" ("the style is the man himself"). The word "style" was not meant as surface gloss, or simply beauty of phrasing. Rather, it was to denote how clearly and coherently meaning could be conveyed through words. In this, he had few equals.

[*] R. L. Williams, ed., *Botanophilia in Eighteenth-Century France: The Spirit of the Enlightenment* (Dordrecht: Kluwer, 2001).

[†] D. Outram, *Georges Cuvier: Vocation, Science and Authority in Post-Revolutionary France* (London: Palgrave Macmillan, 1984).

THE EPOCHS OF NATURE

GEORGES-LOUIS LECLERC,
LE COMTE DE BUFFON

First Discourse

In civil history, one consults documents, studies old medals, deciphers antique inscriptions, to determine the epochs of human revolution and to establish the dates of moral events. Likewise, in natural history, one must rummage through the Earth's archives, pull ancient monuments from the entrails of the Earth, reassemble their remains, and put together in a body of evidence all of the indications of physical change that can allow us to reach back into the different ages of Nature. This is the only way to fix some points in the immensity of space, to place some milestones along the eternal passage of time. Time is like distance: our view would diminish in it, and also even get lost in it, if our history and chronology did not provide lanterns and torches in the darkest places. But, despite the light shed by the written tradition, if one goes back only a few centuries, then what uncertainties in the facts!—what errors are made regarding the causes of events!—and what profound obscurity surrounds yet older times! Moreover, the written word conveys to us only the actions of a few Nations—that is, the actions of a tiny part of the human race. All of the other peoples have left nothing for us, nothing for posterity. They only emerged from nothingness to pass like shadows, leaving no trace—and praise the heavens that the names of all the supposed Heroes, whose crimes and bloody exploits were once celebrated, have passed into oblivion!

Thus, civil history, limited on one side by the shadows of a time that is quite close enough to ours, extends on the other only to small parts of the world, successively inhabited by peoples mindful of their collective memory. Natural history, by contrast, encompasses equally all of space, all of time, and has no limits other than those of the universe.

Nature, being contemporary with matter, space, and time, its history is that of all substances, of all places, of all ages. It may seem at first sight that its great works neither alter nor change, and that, in its productions—even the most fragile and fleeting—it shows itself as forever the same, since at every moment its first models reappear to our eyes in new representations. However, looking more closely, one can see that its course is not completely uniform: one can see that Nature admits distinct variation, shows successive changes, and is even given to forming new combinations and mutations in composition and in form. Thus, as much as it can seem constant overall, it shows variation in all of its parts. And, if we try to encompass its entire span, we cannot doubt that is very different today than it was at its beginnings, and to what it will become as time passes. These are the various changes that we can call its epochs. Nature is found in different states: the surface of the Earth has taken successively different forms: even the heavens have varied, and all things in the physical universe, like those of the moral world, are in a continual movement of successive changes. For example, the state in which we today see Nature is as much our work as its own: we have learned to temper it, modify it, bend it to our needs and desires. We have penetrated deep into, cultivated, and made fertile the Earth: the face it now shows is therefore quite different from that of times before the invention of arts. The golden age of morality—or rather that of fable—was only the iron age of physics and of truth. The people of those days were still half savage, dispersed, few in number, and did not sense their power or know their real worth; the treasure of their lights was still hidden. Man was unaware of the power of a united will and did not suspect that, through society and through sustained, concerted work, he would come to imprint his ideas on the entire face of the universe.

Therefore, one has to go to seek, and look upon, Nature in newly discovered lands, in countries that have remained uninhabited, to gain an idea of its ancient state. And, that ancient state is still modern in comparison to that state where terrestrial continents were covered by the waters, where fish once lived above our plains, where our mountains formed the reefs of the seas. How many changes and different states must have succeeded each other since those antique times (which were not, though, the first times of all) up until the ages of history! How many things lie buried; how many events have been entirely forgotten; and how many revolutions have taken place beyond the reach of human memory! It took a very long succession of observations, over thirty centuries of the culture of the human mind, just to sketch out the present order of things. The Earth has not yet been entirely explored. It is only recently that its true shape has been determined, and it is only in our days that one has been able to form a theory of its internal structure, to demonstrate the order and arrangement of the materials of which it is composed. Hence it is only now that one can begin to compare Nature with itself, and reach back from its known, present-day state into epochs of a more ancient state.

But it is a case here of penetrating the darkness of time: of surveying, by observation of present-day objects, the former existence of things now destroyed, and to get back, through the sole force of substantive facts, to the historical truth of buried things. It is a matter, in a word, of judging, not just the modern past, but also the much more ancient past, only from what is in the present. For us to reach this viewpoint, we need to combine all our powers, and to make use of three main pathways: (1) those facts that allow us to get near the origin of Nature; (2) the monuments that one can regard as witness to its first ages; (3) the traditions that can give us some idea of subsequent ages. After this, we must try to tie together the whole through use of analogies and to form a chain that, from the summit of the ladder of time, can descend as far as to us.

First Fact The Earth is raised at its equator and depressed at its poles, in the proportions demanded by the laws of gravity and of centrifugal force.

Second Fact The terrestrial globe has its own internal heat, which is independent of that conveyed to it by the rays of the Sun.

Third Fact The heat that the Sun sends to the Earth is relatively small by comparison with the heat within the terrestrial globe; and that the heat from the Sun would not be sufficient by itself to sustain living Nature.

Fourth Fact The materials that make up the globe of the Earth are in general of the nature of glass, and can all be reduced to glass.

Fifth Fact One finds on all of the Earth's surface, and even on mountains up to 1,500 to 2,000 toises in height, an immense quantity of shells and other debris of the sea's production.

Let us look first if, in these facts that I wish to use, there is anything that one can reasonably dispute. Let us see if they can all be proved—or at least if they are capable of being so. After that we will pass on to the deductions that we can draw from them.

The first fact, of the bulging out of the Earth at the equator and its flattening at the poles, is mathematically demonstrated and physically proven by the theory of gravitation and by experiments with a pendulum. The terrestrial globe has precisely the shape that a fluid globe would take that rotates with the speed that we know that the Earth's globe possesses. Therefore the first consequence arising from this incontestable fact is that the matter of which our Earth is made was in a fluid state at the moment when it took its form, and that that was also the moment when it first began to rotate. If the Earth had not been fluid, and had it the same consistency that we see in it today, it is clear that the consistent and solid matter would not have obeyed the law of centrifugal force, and therefore, despite the speed of its rotation, the Earth, rather than being a spheroid bulging out at its equator and being flattened at the poles, would be a perfect sphere and it could not have taken any shape but that of a perfect sphere, by virtue of the mutual attraction of all of the particles of matter of which it is composed.

And, as in general all fluidity has heat as its cause, since water itself without heat would form only a solid substance, we have two different manners of appreciating the possibility of this primitive state of fluidity of the terrestrial globe, because it seems initially that Nature has two means of effecting this. The first is of dissolution or even the suspension of terrestrial materials in water, and the second is liquefaction by fire. But we know that the great majority of the solid materials that make up the terrestrial globe are not soluble in water—and at the same time we can see that the amount of water is so small in comparison to dry matter, that it is not possible that one could ever have been suspended in the other. Thus the fluidity, which once was possessed by the entire mass of the Earth, could not have been caused either by dissolution or by suspension in water. That fluidity must have been a liquefaction caused by fire.

This just consequence, already reasonable in itself, gains a new degree of probability through the second fact, and it becomes a certainty from the third fact. The internal heat of the Earth, that is still present and that is much greater than that which comes to us from the Sun, shows us that this ancient fire, which the Earth experienced, has not yet dissipated entirely. The surface of the Earth is colder than the interior. Certain and repeated observations assure us that the entire mass of the Earth has its own heat that is quite independent of that from the Sun. This heat is manifest to us by comparison of our winters and our summers, and one senses it, in a manner that is yet more palpable, once one penetrates below the surface. It is constant in all places for each depth, and seems to increase the farther one descends.[1] But what are such observations in comparison to those that would be needed to trace the successive degrees of internal heat in the depths of the Earth? We have excavated into the mountains to depths of a few hundred toises to extract metals. On the plains, we have dug wells to a few hundred feet. These are our greatest excavations, or rather our deepest trenches. They barely scratch the surface of the outer rind of the Earth, and nevertheless the internal heat there is already more noticeable than that at the surface. One therefore presumes that if one penetrates farther, the heat would become greater, and that those regions close to the center of the Earth are hotter than those more distant from it—just as one sees a fire-heated cannonball maintain its incandescence internally long after its surface has lost its incandescent state and red glow. The fire—or rather the internal heat—of the Earth is also indicated by the effects of electricity, which converts this invisible heat into luminous thunderbolts. It is shown to us by the temperature of the water in the sea, which at the same depths is about equal to that of the interior of the Earth.[2] Besides, it is easy to show that the liquid state of the water in the oceans in general cannot be attributed to the power of the Sun's rays, because it can be shown by experiment that sunlight does not penetrate more than six hundred feet[3] even through the most limpid of waters, and therefore its heat does not reach even a quarter of this depth, that is to say, to one hundred and fifty feet.[4] Thus, all of the waters that lie below this depth would be frozen, without the internal heat of the Earth that can maintain its liquid state. Similarly, it can be proven by experiment that the heat of the Sun's rays do not penetrate more than fifteen or twenty feet into the Earth, because ice can be kept at such depths even during the hottest summers. Thus it is evident that beneath the ocean basins, just as in the primary layers of the Earth, there is a continual emanation of heat that keeps water liquid and gives rise to the temperature of the Earth. Thus there is a heat in its interior that inherently belongs to it, and is quite independent of that which the Sun can provide.

We can confirm this general fact by a great number of specific facts. Everyone has noticed that, in the times of frost, snow melts in all of those places where vapors from the interior of the Earth have free passage, as above wells, covered aqueducts, vaults, cisterns, and so on, while, over the rest of the area where the ground is gripped by ice and intercepts these vapors, snow remains, and freezes instead of melting. That in itself suffices to show that the emanations from the Earth's interior possess a quantity of heat that is very real and detectable. But it is useless here to wish to gather new proofs of a fact that is clear from experiment and observation. It is sufficient that one cannot henceforth place this into doubt, and that one recognizes the internal heat of the Earth as a real and genuine fact, from which, as from other general facts of Nature, one must deduce particular consequences.

There is from this a fourth fact: one cannot doubt, following demonstrable proofs that we have given in several sections of our *Theory of the Earth*, that the materials of which the Earth is made are of the nature of glass.[5] The basis of minerals, of plants and animals, is rather made of vitrescible matter, because all of their residues, all of their subsequent waste, can be reduced to glass. The materials that the chemists call *refractory* and those that they regard as infusible, because they resist the fires of their furnaces, can nevertheless be so reduced by the action of a more intense fire. In this way, all of the materials that compose the globe of the Earth—at least all that are known to us—have glass as the basis of their substance,[6] and we can, on submitting them to the powerful action of fire, subsequently reduce them all to their primary state.

The primitive liquefaction of the entire mass of the Earth by fire is therefore proved with all the rigor demanded by the strictest logic. First, a priori, by the first fact of the Earth's own elevation at its equator, and its depression at the poles. Second, *ab actu*, by the second and third facts concerning the Earth's still-persistent internal heat. Third, a posteriori, by the fourth fact, which shows us the product of this action of fire, that is to say, glass, in all terrestrial substances.

But although the materials that make up the globe of the Earth were primordially of the nature of glass, and one can reduce them to this subsequently, one must nevertheless distinguish and separate them, relative to the different states in which they are found before they return to their primary state, that is to say, before they revert to glass through the action of fire. This consideration is all the more necessary here, because only that can show us how the formation of these materials differs. One thus first has to divide them into vitrescible materials and calcinable materials. The first of these do not show any effect from fire unless they are subjected to a sufficient degree of its force to convert them into glass. The others, by contrast, undergo, at much lesser levels, a process that converts them into lime. The quantity of calcareous substances, although considerable on Earth, is nevertheless very small by comparison with the quantity of the vitrescible materials.

The fifth fact we have put forward proves that their formation also came at a different time and from a different element. And one sees clearly that all materials that were not formed directly by the action of primitive fire were formed via the intermediary of water—because all are made of shells and other debris that are the products of the sea. We place in the class of vitrescible materials quartz, crystalline rock, sands, sandstones, and granites; also slates, shales, clays, metals, and metallic minerals. These materials taken together form the true foundations of the globe and make up the principal and by far the greater part of it.

All were originally produced by primitive fire. Sand is only powdered glass, and clays are sands that have decayed in water. Slates and shales are dried and hardened clays. Quartz, crystalline rock, sandstones, and granite are only vitreous masses or vitrescible sands in concrete form; pebbles, crystals, metals, and most other minerals are only distillates, exudations, or sublimates of primary materials, which all reveal to us their primitive origin and common nature, by their aptitude to be reduced directly into glass.

But the calcareous sands and gravels, chalks, freestones, rubble stones, marbles, alabasters, calcite spar, both opaque and transparent—all the materials that, in a word, transform into lime, do not at first show their primary nature. Although originally glassy like all the others, these calcareous materials have passed along pathways that have denatured them. They have been formed in water: all are entirely composed of shells and of the detritus of those aquatic animals that alone know how to convert liquid into solid and transform seawater into stone. Common marbles and other calcareous rocks are composed of entire shells and fragments of shells, of corals, of starfish, and others of which the parts are still evident or easily recognizable. Calcareous gravels are only the debris of marbles and limestones that the action of air and freezing has detached from rocky crags, and one can make lime out of such gravels, out of marble or out of rock; one can do the same with the shells themselves, and with chalk and with tufas, which are yet only debris or rather decayed remains of the same substances. Alabasters and comparable marbles that contain alabaster can be recognized as large stalactites, which form at the expense of other marbles and out of common rocks; the calcite spars form similarly by exudation or distillation in calcareous materials, just as rock crystal forms in vitrescible materials. All this can be proved by inspection of these materials and by close examination of the monuments of Nature.

First Monuments One finds shells and other marine products on the surface and in the interior of the Earth; all of the materials that one can call limestone are made out of their remains.

Second Monuments On examining the shells and other marine products that one can extract from the ground in France, Germany, and the rest of Europe, one sees that a large proportion of the animal species to which these remains belonged are not found in the adjacent seas, and these species either no longer survive, or are found only in more southerly seas. Similarly, one finds in slates and in other such materials, at great depths, impressions of fish and plants, none of which can be found in our climes, and these either exist no longer, or are found in southern climes.

Third Monuments One finds in Siberia and in other northern countries of Europe and Asia, skeletons, tusks, and bones of elephants, hippopotami, and rhinoceri, in large enough quantities to be assured that species of these animals, which today can propagate only in the lands of the south, existed and propagated previously in the northern lands. One can see that these remains of elephants and other terrestrial animals occur at shallow depths, while shells and other debris of marine production are found buried at greater depths in the interior of the Earth.

Fourth Monuments One finds elephant tusks and bones and also the hippopotamus teeth, not only in the northern lands of our continent, but also in those of North America, while species of elephant and hippopotamus do not now exist at all in the continent of the New World.

Fifth Monuments One finds in the interiors of continents, in regions that are the farthest removed from the sea, an infinite number of shells, of which most belong to animals currently living in southern seas, and of which a number of others have no living analogue, and so these species seem to be lost and destroyed, through causes that are at present unknown.

In comparing these monuments with the facts, one sees first that the time of formation of vitrescible minerals is considerably more distant than that of calcareous substances. And, it seems that we can already distinguish four and even five epochs in the greatest depths of time. The first: when the matter of the globe was in a state of fusion through fire, and the Earth took its shape and was elevated at the equator and depressed at the poles by its movement of rotation. The second: when the matter of the Earth, being solidified, formed great masses of vitrescible material. The third, when the sea, covering a land presently inhabited, nourished shell-making animals of which the remains formed calcareous substances. And the fourth: when the retreat took place of the seas that had covered the continents. A fifth epoch, as clearly marked as the first four, is that of the times when the elephants, hippopotami, and other animals of the south lived in the northern regions. This epoch is evidently later than the fourth, because the remains of these terrestrial animals are found almost at the ground surface, while those of the marine animals are, in the same regions, mostly found at great depths.

What?—one might say. Did the elephants and other animals of the south formerly inhabit the lands of the north? This singular fact, as extraordinary as it may seem, is no less certain for that. People have found and find still, all the time, in Siberia, in Russia, and in other northern countries of Europe and Asia, ivory in great quantities. These elephant tusks may be extracted from a few feet below the ground, or are disinterred by running water as it undercuts the ground at river banks. One finds these bones and tusks of elephants in so many different places and in such abundance that one can no longer say that they are the remains of a few elephants transported by humans into cold regions. One is now forced to say by these repeated proofs that these animals naturally inhabited the countries of the north, just as they live today in southern lands. And, what seems to render this fact yet more marvelous—that is to say, more difficult to explain—is that one finds these southern animal remains on our own continent, not only in our northern regions, but also in the lands of Canada and other parts of North America. We have in the Cabinet du Roi several tusks and a great number of bones of elephants that were found in Siberia; we have other tusks and other elephant bones that have been found in France, and finally we have the elephant tusks and hippopotamus teeth found in America in the land bordering the Ohio River. It is thus unavoidable that these animals which can—and do—only exist today in hot countries, formerly lived in northern climes and so, by consequence, this cold zone must have been as hot then as is our torrid zone today. For, it is not possible that the constitutive form, or, if one wishes, the true habitus of animal bodies, which is one of the most fixed things in nature, can change to the point of giving the temperament of the reindeer to the elephant, nor to suppose that the animals of the south, which need great warmth to survive, could live and multiply in the northern regions, if the climate had been as cold then as it is today. M. Gmelin, who has traversed Siberia and himself collected a number of elephant bones in these northern regions, seeks to explain this by proposing that the great inundations that affected southern regions drove the elephants to the northern countries, where they all perished at once because of the harshness of the climate. But this proposed cause is not proportional to the effect. There has perhaps already been more ivory found in the north than all of the elephants of India living today could provide. One will find much more ivory over time, when the vast deserted regions of the north, which are barely explored, will be inhabited, and when these regions will be shaped and dug over by the hand of man. Besides, it seems decidedly strange that these animals took the route that conformed least to their nature, because if one supposes that they were pushed out of the south by the floods, there would have been two natural escape routes—to the east and to the west. And why flee as far as sixty degrees north while they could stop en route and go off to the side toward happier regions? And how can we conceive that, due to a flood from southern seas, they could have been chased a thousand leagues into our continent and more than three thousand leagues into the other? It is impossible that a marine inundation in the Far East would have sent elephants to Canada or even Siberia, and it is equally impossible that they would arrive in numbers as large as indicated by their remains.

Being little satisfied by such an explanation, I have thought that one could give another explanation that is more plausible and that accords perfectly with my theory of the Earth. But before presenting this, I will observe, to forestall any difficulties: (1) That the ivory that one finds in Siberia and in Canada is certainly elephant ivory, and not the ivory of a walrus or sea cow, as some travelers have suggested: one also finds fossil walrus ivory in southern countries, but it is different from that of an elephant, and it is easy to distinguish the bones by comparing their internal texture. The tusks, the molar teeth, the shoulder blades, the femurs, and other bones found in the northern regions are certainly elephant bones. We have compared them to the various respective parts of an entire elephant skeleton, and one cannot doubt their identity as species. The large square teeth found in the same northern lands, whose biting surface is clover-shaped, have all the characters of the molar teeth of a hippopotamus; and the other enormous teeth, whose biting surface is made up of large rounded points, belonged to a species now obliterated on the Earth, just as the large spirals termed the horns of Ammon are now obliterated in the sea.

(2) The bones and tusks of these ancient elephants were at least as large and as thick as those of today's elephants[7] that we have made comparison with. This proves that these animals were not forced to live in these regions, but lived there in a state of Nature and in complete liberty, because they attained there their largest dimensions and fullest growth. Thus one could not presume that they were transported there by humans. Just in itself, the state of captivity, quite apart from the rigors of the climate,[8] would have reduced them to a quarter or a third of the size that is revealed by their remains.

(3) That this great quantity of remains has already been found by chance in near-deserted regions, where no one looks for them, suffices to show that this is not by one or a few accidents, nor on a single occasion, when a few individuals of that species happened to be found in the northern lands, but it is an absolute necessity that these very species lived, fed, and multiplied there, just as they live, feed, and multiply today in the countries of the south.

Put like this, it seems to me that the question comes down to knowing, or rather consists of looking if there is or was, a mechanism that could change the temperature in different parts of the globe, to the point at which the lands of the north, today very cold, could have previously been as warm as the southern lands. Some physicists are given to think that this effect can be produced by a change in the obliquity of the ecliptic: because, at first sight, this change seems to show that the inclination of the axis of the globe not being constant, the Earth could have previously turned around an axis quite distant to that around which it turns today, so that Siberia could have found itself beneath the equator. The astronomers have observed that the change in the obliquity of the ecliptic is of the order of forty-five seconds every century. Thus, supposing that the increase is progressive and constant, it needs only sixty centuries to produce a difference of forty-five minutes, and three thousand and six hundred centuries to give one of forty-five degrees—that which would bring the sixtieth degree of latitude to the fifteenth, that is to say, the Siberian regions, where elephants once lived, to the region of India, where they live today. So it is only a matter, one says, of stating that there was such a long period of time in the past, to explain the presence of elephants in Siberia: three hundred and sixty thousand years ago, the Earth turned on an axis aligned forty-five degrees away from that around which it turns today; the fifteenth degree of latitude is now the sixtieth, and so forth.

To this, I reply that this idea and means of explanation cannot be upheld, once one comes to examine them. The change in the obliquity of the ecliptic is not a decrease or increase that is progressive and constant. To the contrary, it is a limited variation, at times in one direction and at times in the other, and so consequently could never produce in any fashion or for any climate a difference of forty-five degrees in inclination: because, the variation in the obliquity of the axis of the Earth is produced by the planets, which displace the ecliptic without moving the equator. Taking the most powerful of these attractions, that of Venus, it would take twelve hundred and sixty thousand years to change by one hundred and eighty degrees the situation of the ecliptic on the orbit of Venus—and by consequence produce a change of six degrees and forty-seven minutes in the real obliquity of the axis of the Earth, since six degrees and forty-seven minutes is twice the inclination of the orbit of Venus. Similarly, the action of Jupiter could only, in nine hundred and thirty-six thousand years, change the obliquity of the ecliptic by two degrees and thirty-eight minutes, and then this effect is in part compensated by the preceding effect. And thus, it is not possible that change in the obliquity of the Earth's axis could ever reach six degrees, unless one supposes that all the orbits of the planets will themselves change—a supposition that we cannot, or should not, admit because there is no mechanism that could produce this effect. And as we can determine the past only by observation of the present and by our view of the future, it is not possible, however far one wishes to push back the limits of time, to suppose that the variation in the ecliptic could ever have produced a difference of more than six degrees in the climates of the Earth—and so this mechanism is quite insufficient, and the explanation that one would wish to draw from it has to be rejected.

But I can provide this so-difficult explanation and deduce it from immediate causes. We have seen that the terrestrial globe, as it took its form, was in a fluid state, and it can be shown that water could not have caused the dissolution of terrestrial material, and so this liquid state must have been caused by fire. And, to pass from this primary state of heating and liquefaction to a gentle and temperate warmth, time was needed. The Earth could not cool suddenly to the state that it is in today. Thus, in the times just after its formation, the Earth's own heat was infinitely greater than that which it received from the Sun, since it is still greater today. Thus, this great fire being dissipated little by little, the climate of the poles underwent, as did all the other climates, successive stages of diminishing heat and of cooling. There was hence a time, and even a long period of time, when the northern regions, having roasted as did all the other regions, enjoyed the same warmth that is present today in the southern regions. Therefore, these northern lands could be and were inhabited by those animals that now live in the south, and that have need of such warmth. Hence that fact, far from being astonishing, ties in perfectly with the other facts, and is only a simple consequence. Rather than contradicting the theory of the Earth that we have now established, this same fact rather becomes additional proof, which can only confirm it where it is most unclear, that is to say, where one begins to fall into such a depth of time that the light of the imagination appears to be extinguished and where, for lack of evidence, it seems to be unable to guide us further.

A sixth epoch, subsequent to the other five, is that of the separation of the two continents. It is certain that they were not separated when the elephants also lived in the northern regions of America, of Europe, and of Asia. I say this also because one finds their bones in Siberia, in Russia, and in Canada. The separation of the continents thus happened only in the times after those animals lived in the northern lands. But, as one also finds elephant tusks in Poland, in Germany, in France, and Italy,[9] one must conclude that as those northern regions cooled, these animals retreated toward the temperate zones where the heat of the Sun and the greater thickness of the globe compensated for the loss of the Earth's internal heat. And, when those zones finally also became too cold over time, they then reached the climates of the torrid zone, which are those where internal heat is conserved the longest by the greater thickness of the spheroid of the Earth, and the only ones where this heat, combined with that of the Sun, is still strong enough today to maintain them, and support their propagation.

Similarly, there are found, in France and all other parts of Europe, shells, skeletons, and vertebrae of marine animals that can only live in the most southerly seas. It was therefore the case that the climates of the seas showed the same changes in temperature as did those of the land; and, as this second fact may be explained, like the first one, by the same causes, this seems to confirm the point of this thesis.

As one compares the ancient monuments of the first age of a living Nature with today's forms, one clearly sees that the constitutive form of each animal has been similarly conserved and without alteration of its principal parts. The type of each species has not changed; the inner mold has conserved its form and has not varied. However long one imagines the passage of time might be, and however many generations one admits or supposes, the individuals of each kind today represent the forms of those first centuries, especially regarding the *espèces majeures*, of which the imprint is most distinct and the nature most fixed. This is because the *espèces inférieures* have, as we have said, visibly undergone all of the effects of different types of degeneration. Only, one can remark regarding the *espèces majeures*, such as elephant and hippopotamus, that in comparing their remains with those of our days, one sees in general that these animals were larger than they are today. Nature was in its first flush of vigor; the internal heat of the Earth gave its predecessors all of the force and the extent of which they were susceptible. There were, in this first age, giants of all kinds: the dwarfs and pygmies arrived later, that is to say, after the cooling and if (as other of the monuments seem to indicate) there were species that were lost—thus, animals that had once existed and that no longer exist—they could be only those whose nature demanded a warmth greater than that of today's torrid zone. Those enormous molar teeth, almost square and with great rounded points, those great petrified spirals, some of which are several feet in diameter,[10] and a number of other fish and fossil shells of which no living analogues can be found anywhere, these existed in those first times only, when the land and the sea were still hot and had to nourish these animals for which this degree of heat was necessary—and which no longer exist, probably since they perished because of the cooling.

There, thus, is the order of time indicated by the facts and the monuments. There, we have six epochs in the succession of the first ages of Nature: six spans of time of which the limits, although undetermined, are no less real, because these epochs are not like those of civil history, marked by fixed points, or limited by centuries and other segments of time that we can count and measure exactly; nevertheless, we can make comparisons among them, and evaluate their relative duration, and relate other monuments and facts to each of these time intervals, which will reveal their contemporary dates, and maybe also other intermediate and subsequent epochs.

But before going further, we must make haste to anticipate a grave objection, which could even degenerate into an imputation. How could you reconcile, one might say, such a great age that you give to matter, with the sacred traditions, which provide for the world only some six or eight thousand years? However strong might be your proofs, however well-founded might be your reasoning, however evident might be your facts—are not those reported in the Holy Book yet more certain? To contradict them, is that not to lack respect to God, who had the goodness to reveal them to us?

I am grieved each time that one abuses the great, the holy Name of God; I am hurt each time that people profane Him, and prostitute the idea of the first Being, in substituting for it the phantoms of their opinions. The more I have penetrated into the heart of Nature, the more I have admired and profoundly respected its Creator. But, a blind respect would be superstition; true religion entails by contrast an enlightened respect. Let us see then: let us try to hear reasonably the first facts that the divine Interpreter has transmitted to us on the subject of the creation; let us collect with care the rays that have escaped from that first light; far from obscuring the truth, they can only add a new degree of luminosity and of splendor.

IN THE BEGINNING GOD CREATED THE HEAVENS AND THE EARTH.

That does not mean to say that at the beginning God created the heavens and the Earth *as they are now*, because it is said immediately after *that the Earth was formless*; and that the Sun, the Moon, and the stars were placed in the heavens only on the fourth day of creation. One would make the text contradict itself if one wished to uphold that *at the beginning God created the Earth and the heavens as they are now*. It was at a subsequent time that He made them in effect *as they are*, giving form to matter, and placing the Sun, the Moon, and the stars in the heavens. Thus to here reasonably understand those words, it is necessary to supply a word that reconciles the whole, and read:

At the beginning God created THE MATTER of the heavens and the Earth.

And *this beginning*, the first time, the most ancient of all, during which the matter of the heavens and the Earth existed without a determinate form, seems to have had a long duration; because let us carefully listen to the words of divine Interpretation.

THE EARTH WAS FORMLESS AND ENTIRELY NAKED AND DARKNESS COVERED THE FACE OF THE DEEP, WHILE A WIND FROM GOD SWEPT OVER THE FACE OF THE WATERS.

The Earth *was*, darkness *covered*, the spirit of God *was*. These expressions, using the past tense of the verb—do they not indicate that it was during a long interval that the Earth was formless and that darkness covered the face of the abyss? If this formless state, this dark face of the abyss, had existed only for a day, or even if this state did not last long, the sacred Writer would either have expressed this differently, or would have made no mention of this moment of darkness. He would have passed from the creation of matter in general to the production of its particular forms, and would not have made a distinct rest, a marked pause between the first and the second instant of the works of God. I therefore clearly see that not only one can, but even that one must, to conform to the sense of the text of the holy Scripture, regard the creation of matter in general as more ancient than the particular and successive productions of its particular forms; and this is further confirmed by the following transition:

THEN GOD SAID

This word *then* suggests things that have been made and things still to make; it is the plan of a new design, it is the indication of a command to change ancient or existing things into a new order.

LET THERE BE LIGHT, AND THERE WAS LIGHT.

Here is the first word of God; it is so sublime and so swift that it shows us well enough that light was produced in an instant, though the light did not appear first nor suddenly as a universal flash, it was for a while combined with darkness, and God himself took the time to consider this, for it is said:

AND GOD SAW THAT THE LIGHT WAS GOOD, AND GOD SEPARATED THE LIGHT FROM THE DARKNESS.

The act of separation of the light from the darkness is therefore distinct and physically separated by a period of time from the act of its production, and this time, during which it pleased God to consider it to see that *it was good*, that is to say useful for its purpose.

This time, I say, still belongs to and should be joined with that of the chaos, which only began to become ordered when light was separated from darkness.

Here there are thus two spans, here two intervals of duration that the sacred Text forces us to recognize. The first, between the creation of matter in general and the production of light. The second, between the production of light and its separation from darkness. In this way, far from lacking respect to God and in giving matter a greater age than the Earth *as it is*, it by contrast shows as much respect as is within us, in conforming our intelligence to his Word. In effect, does not the light that illuminates our souls come from God? Can the truths that it shows be contradictory to those that He has revealed to us? We must remember that this divine inspiration has been passed through human organs; that His words have been transmitted to us in an impoverished language, bereft of precise expressions for abstract ideas, meaning that the Interpreter of this divine word had been obliged to frequently employ words of which the understanding is determined only by circumstances. For example, the words *to create* and the words *to form* or *to make* are used indeterminately to signify the same thing or similar things, while in our language these two words each have a very different and very determinate sense: "to create" is to draw something out of nothingness, "to form" or "to make" is to draw it out of something into a new form; and it seems that the words "to create" belong by preference and perhaps uniquely to the first verse of Genesis, of which the precise translation in our language should be *at the beginning God drew out of nothingness the matter of the heavens and of the earth*; and that which proves that the words *to create* or *to draw out of nothingness* can be applied only to these first words, is that all of the matter of the heavens and of the Earth was created or drawn out of nothingness from the beginning. By consequence, it is no longer possible or permissible to consider new creations of matter, since then *all matter* would no longer have been created from the beginning. Consequently the work of six days can be understood only as a formation, a production of forms drawn from matter created previously, and not as other creations of matter drawn directly out of nothingness. And, in effect, while it is a question of the light that is the first of these formations or productions drawn out of the heart of matter, it is said only *that the light was made* and not *that the light was created*. All goes together, thus, to prove that matter was created *in principio*, and it was only in subsequent times that it pleased the sovereign Being to give it form, and in the place of creating and forming everything in the same instant, as He could have done, if He had wished to deploy the full extent of his Omnipotence, He wished only, by contrast, to act over time, to produce successively and even place periods of repose, considerable intervals, between each of his works. What can we understand by the six days that the sacred Writer shows us so precisely in counting them one after the other, if not six spans of time, six intervals of duration? And these spans of time indicated by the name of *days*, for lack of other expressions, cannot have any correspondence with our current days, because three of these passed before the Sun had been placed in the sky. It is thus not possible that these days were like ours, and the Interpreter of God seems to show this well enough by always counting from the evening until the morning, rather than as solar days that should be counted from morning to evening. These six days were therefore not solar days like ours, nor even days of light, because they start in the evening and finish in the morning. These days were not even equal, because they were not proportional to the work done. They were thus only six spans of time. The sacred Historian did not establish the duration of each, but the sense of the narration seems to make them reasonably long, so that we can extend them to the duration demanded by the *physical* truths that we have to show. Why then protest so loudly on this use of time that we can only make as much of as we are forced to by our demonstrable knowledge of the phenomena of Nature? Why wish us to refuse this time span, since God gives it to us by His own words and that it would be contradictory or unintelligible if we did not admit the existence of this first time before the formation of the Earth *such as it is*?

It is fine that one can say, that one can uphold, even rigorously, that since the latest phase, since the end of the works of God, that is to say since the creation of man, there has been no more than six or eight thousand years, because the different genealogies of the human kind since Adam do not indicate more. We own this belief, this mark of submission and of respect to the most ancient, the most sacred of all traditions; we owe it yet more, that it can never allow us to deviate from the letter of the sacred tradition except when *the letter kills*, that is to say, when it seems directly opposed to healthy reason and to the truth of the facts of Nature. Because all reason, all truth, comes equally from God, there is no difference between the truth that He has revealed and the truth that we are allowed to discover by our observations and our researches. There is, I say, no other difference than that between a first favor made freely and a second that He has wished to separate and make us deserve through our work. And, it is for this reason that His Interpreter talked to the first men, who were still very ignorant, only in a common sense. It was not raised above their understanding which, far from reaching toward the true system of the Earth, did not even go beyond commonly held notions, found on what their senses told them. Because, in effect, it was to those people that it was necessary to speak, and this speech would have been in vain and unintelligible if it had been in a form that one could state today, since even today there are only a small number of men to whom the astronomical and physical truths are known well enough to not allow them to be doubted, and who are able to hear their language.

Let us see thus what Physics was in these first ages of the world, and what it would still be if man had never studied Nature. One sees the heavens as a blue arch in which the Sun and the Moon seem to be the greatest stars, of which the first always produces the light of day and the second often makes that of the night; one sees them appear or rise on our side and disappear or set on the other, after having run their course and given their light during a certain interval of time. One sees that the sea is of the same color as that blue arch, and that it seems to touch the heavens, when one sees it from afar. All people's ideas on the system of the world carry only toward those three or four notions; and however false these may be, it is necessary to conform to them in order to make ourselves understood.

Because the sea seems in the distance to unite with the heavens, it was natural to imagine that there exist higher and lower waters, of which one fills the heavens and the other fills the sea, and that to hold up the waters above them there is need of a firmament, that is to say, a support, a solid and transparent arch, through which one can see the blue of the waters overhead; thus it is said: *That the firmament be made in the middle of the waters and that it separates the waters from the waters; and God made the firmament, and separated off the waters that were above the firmament from those that were below the firmament, and God gave to the firmament, the name of the heavens . . . and to those waters gathered under the firmament, the name of the sea.* It is to these same ideas that are ascribed the cataracts of the heavens, that is to say, the doors or the windows of the heavens that open when it is necessary to allow the waters from above to drown the earth. It is also from these same ideas that it is said that the fish and the birds have a common origin. The fish would have been produced by the lower waters, and the birds by the waters above, because they approached in their flight the blue dome, which the common people did not imagine to be much higher than the clouds. Likewise the people have always thought that the stars are attached like nails to this solid dome, that they are smaller than the Moon and infinitely smaller than the Sun; they did not even distinguish the planets from the fixed stars; and it is for this reason that no mention is made of the planets in all of the story of creation; it is for this reason that the Moon is regarded in it as the second star, although it is in effect only the smallest of all the celestial bodies—and so on, and so on.

Everything in the story of Moses is placed within the limits of intelligence of the people. Everything there is represented relative to the common man, to whom it would not do to demonstrate the true system of the Earth, but it was sufficient to instruct about that which he owed to the Creator, in showing him the effects of His omnipotence as so many good deeds. The truths of Nature were to appear only after time, and the sovereign Being kept them to Himself as the surest means of recalling man to Him, as his faith, declining over centuries, would become unsteady; far from his origins, man could forget Him. Finally, too accustomed to the spectacle of Nature, man was no longer moved by it and came to disregard its Author. It was therefore necessary to reaffirm from time to time, and even to enlarge, the idea of God in the spirit and in the heart of man. And each discovery produced this great effect: each new step we make into Nature brings us closer to the Creator. A newly seen truth is a kind of miracle: the effect is the same, and it differs from a true miracle only in that it is a sudden bolt through which God strikes immediately and rarely. Instead He serves man to discover and demonstrate the marvels with which He has filled the heart of Nature. As these marvels operate at each instant, and as they are exposed always and for everybody for their contemplation, God calls it constantly to man's mind, not only by the spectacle itself but also by the successive development of these works. For the rest, I permit myself this interpretation of the first verses of Genesis only with the view of bringing about a great good: this would be a reconciliation forever between the science of Nature and that of theology. They can only, according to me, be in apparent contradiction; and my explanation seems to demonstrate this. But if this explanation, however simple and very clear, seems insufficient and even out of the question to certain minds who are too strictly attached to the letter, I ask them to judge me by my intention, and to consider that my system of the Epochs of Nature, being purely hypothetical, cannot hence harm the revealed truth, which comprises so many immutable axioms, independent of all hypothesis, to which I have submitted and submit my thoughts.

FIRST EPOCH

When the Earth and the Planets Took Their Form

In this first time, when the molten Earth, turning around itself, took its form and became elevated at the equator and depressed at the poles, the other planets were in the same state of liquefaction, because in turning around themselves they took, like the Earth, a form expanded around their equator and flattened at their poles, and this expansion and flattening is proportional to the speed of their rotation. The globe of Jupiter provides us with proof: as it spins much more rapidly than that of the Earth, it is by consequence much more elevated at its equator and more depressed at its poles, because observations show us that the two diameters of this planet differ by more than one-thirteenth, while those of the Earth differ only by one part in 230. They show us this also with Mars, which spins at less than half the speed of Earth, and the difference between the two diameters is too small to be measured by astronomers. And, on the Moon, of which the movement of rotation is yet slower, the two diameters seem equal. The speed of rotation of the planets is thus the sole cause of their expansion at the equator, and this expansion, made at the same time as their flattening at the poles, suggests complete fluidity in all of the mass of these globes, that is to say, a state of liquefaction caused by fire.*

Moreover, all of the planets circle around the Sun in the same direction, and nearly all in the same plane. They seem to have been set in motion by a common impulsion, and at the same time. Their movement of circulation and their movement of rotation are contemporaneous, just as well as their state of fusion or their liquefaction by fire, and these movements had necessarily been preceded by the impulsion that produced them.

In those planets of which the mass had been struck the most obliquely, the movement of rotation was the most rapid; and by this rapidity of rotation, the first effects of centrifugal forces exceeded those of weight; in conse-

* See *Theory of the Earth*, article on the formation of the planets, vol. 1 of *Histoire naturelle*.

quence there took place in these liquid masses a separation and a projection of parts to the equator, where this centrifugal force is greatest, and those parts separated and thrown out by this force formed concomitant masses, formed satellites, that had to circulate and that all circulated, in effect, in the plane of the equator of the planet from which they had been separated by this cause. The satellites of the planets were thus formed at the expense of the matter of their principal planet, as the planets themselves seem to have formed at the expense of the mass of the Sun. Thus the time of formation of the satellites is the same as that when the rotation of the planets began. This is the moment when the material that makes them up was assembled and still formed only liquid globes, a state in which this liquefied material could be separated and easily projected outward; because, as soon as the surfaces of these globes had begun to acquire a little consistency and rigidity from cooling, the material, although animated by the same centrifugal forces, being held by the forces of cohesion, could no longer be separated nor projected beyond the planet by the same movement of rotation.

As we do not know in Nature any cause of heat, any fire other than that of the Sun, which could melt or keep in liquefaction the matter of the Earth and of the planets, it seems to one that in refusing to believe that the planets have been derived from the Sun, one would be at least forced to suppose that they had been at least exposed from very close range to the heat of that fiery star, to be liquefied. But this supposition would not yet be sufficient to explain the effect, and would fail of its own accord, through an unavoidable circumstance: for it would need time for the fire, however fierce it would be, to penetrate the solid matters exposed to it, and a very long time to liquefy them. One has seen by the preceding experiments, that to heat a body to the point of fusion, one needs at least a fifteenth of the time that is needed to cool it, and seeing the great volume of the Earth and of the other planets, they would have needed to be stationary for thousands of years by the Sun to receive the degree of heat necessary to liquefy them: and, there is no example in the universe of any body, any planet, any comet resting stationary by the Sun, even for an instant. By contrast, the closer that comets approach the Sun, the more rapid is their movement; the time of their perihelion is extremely short, and the fire of this star, in burning its surface, does not have the time to penetrate the mass of the comets that approach the closest to it.

Thus, everything combines to prove that it is not sufficient for the Earth and the planets to have passed, like certain comets, into proximity with the Sun so that the liquefaction could take place: we can thus presume that this planetary material once belonged to the body of the Sun itself, and was separated from it, as we have said, by a single and the same impulsion. Because, the comets that approach most closely to the Sun show us only the first degree of the effects of its heat. They seem to be preceded by an inflamed vapor as they approach, and to be followed by a similar vapor as they distance themselves from this star; thus a part of the superficial matter of the comet extends around it and is shown to our eyes in the form of luminous vapors, which are in a state of expansion and volatility caused by the fire of the Sun. But the core,[1] that is to say, the very body of the comet, does not seem to be profoundly penetrated by the heat, because it is not inherently luminous, as would be all of a mass of iron, or glass, or of some other solid matter intimately penetrated by this element. By consequence, it seems necessary that the matter of the Earth and of the planets, which were once in a state of liquefaction, belonged to the very body of the Sun, and that they were part of these substances in fusion that make up the mass of this star of fire.

The planets were given their motion by the single and same impulsion, because they all circulate in the same sense and almost in the same plane. The comets, by contrast, which circulate like the planets around the Sun, but in different directions and planes, seem to have been set into motion by different impulsions. One can thus assign the motion of the planets to a single epoch, while that of the comets could have been given at a different time. Thus nothing can clarify for us the origin of the movement of the comets; but we can apply reason to that of the planets, because they possess common links that show clearly enough that they were set into motion by one and the same impulsion. It is thus allowable to seek in Nature the cause that could have produced such a great impulsion; by contrast, we can scarcely reason upon nor even research into the causes of the movement of impulsion of the comets.

Gathering solely the fleeting evidence and slight indications that can provide us with a few conjectures, one can imagine—to satisfy, however imperfectly, the mind's curiosity—that the comets of our solar system were formed by the explosion of a fixed star or of a neighboring Sun, of which all of the dispersed parts, no longer having a center or a common focus, would have been forced to obey the force of attraction of our Sun, which from then on would have become the pivot and the focus of all our comets. We, and posterity, cannot say further, until when, by further observations, one can come to be able to recognize some common relationship in the movement of impulsion of the comets. For, because we know nothing except by comparison, since we lack any information and no analogy presents itself, all light flees, and not only our reason, but even our imagination, finds itself at a loss. Thus, having previously abstained from making conjectures upon the causes of the impulsive motion that propelled the comets, I have thought myself obliged to think on the impulsion of the planets. I have put forward, not as true and certain fact, but just as a possibility, that the matter of the planets was projected out of the Sun by the shock of a comet. This hypothesis is founded on the absence in Nature of bodies, other than comets, that can or could transmit such a large movement to such large masses, and also on the fact that comets approach the Sun so closely that it is inevitable that some fall in obliquely, and on furrowing the surface, throw material that has been put into movement by the impact out before them.

A similar cause could have produced the heat of the Sun. It seemed to me that one can deduce this from natural effects, and find it in the constitution of the Earth system. Because the Sun, having to support all the weight, all of the action of the penetrating force of the vast bodies that circulate around it, and having to suffer at the same time the rapid action of this kind of internal friction throughout its mass, the material composing it must be in a state of the greatest division. It must have become and remained fluid, luminous, and burning, because of this pressure and this internal friction, always equally active. The irregular movements of the sunspots, such as their spontaneous appearances and disappearances, show well enough that this star is liquid, and that from time to time it lifts above its surface types of scoria or froth, of which some float irregularly above this molten matter, and others are fixed for a time, and then disappear like the first type when the action of the fire has again divided them. We know that it is by means of some of those fixed blotches that we can determine the period of the rotation of the Sun as twenty-five-and-a-half days. Each comet and each planet forms a wheel, of which the spokes are the rays of the attractive force, and the Sun is the axle or the common pivot of all these different wheels. A comet or a planet is a mobile rim and each contributes with all of its weight and all of its speed to the heating of this common axle, of which the fires will consequently last just as long as will the movement and pressure of the large bodies that produce them.

From that, does not one then presume that if one does not see planets around the fixed stars, this is only because of their immense distance from us? Our view is too limited and our instruments too weak to see these dim bodies, because even those that are luminous escape our view, and of the infinite number of those stars, we will ever know only those that our telescopes can make closer to us. But analogy shows us that being fixed and luminous like the Sun, the stars had to be heated, to liquefy and burn by the same cause, that is to say, the active pressure of opaque, solid, and obscure bodies that circulate around them. That alone can explain why only the fixed stars are luminous, and why in the solar universe, all of the wandering stars are obscure.

And, the heat being produced by this cause having to be by reason of the number, the speed, and the mass of the bodies that circulate around the common focus, the heat of the Sun must be of an extreme intensity, or rather an extreme violence, not only because the bodies that circulate around it are all vast, solid, and turn rapidly, but further, because they are great in number. Because, apart from the six planets, the ten satellites, and the rings of Saturn, which all weigh upon the Sun, and make up a volume of material two thousand times greater than that of the Earth, the number of comets is greater than is commonly thought; they alone could suffice to light the fire of the Sun before the flinging out of the planets, and even suffice to maintain it today. Man will maybe never be able to explore the planets that circulate around the fixed stars. But, with time, he could get to know the number of comets in the solar system; I regard this great knowledge as reserved for posterity. While waiting, here is a type of evaluation which, though far from being precise, can help to fix ideas on the number of these bodies circulating around the Sun.

In consulting the observational archives one can see that, since the year 1101 up to 1766, that is, in 665 years, there have been 228 comet sightings. But the number of these errant bodies that have been observed is not as great as the number of appearances, because most if not all complete their revolution in less than 665 years. Let us take thus the two comets of which the revolutions are perfectly known to us: the comet of 1680, of which the period is around 575 years, and that of 1759, of which the period is 76 years. One can believe, in waiting for better evidence, that in taking the average time, 326 years, between these two periods of revolution, there are as many comets of which the period is greater than 326 years as there are of those of which the period is lesser. Therefore in reducing all to 326 years, each comet would have appeared twice in 652 years, and one would have thus about 150 comets for 228 appearances in 665 years.

Now, if we consider that there are likely more comets beyond our field of view, or that escaped the gaze of observers, this number will maybe triple. From this, one can reasonably think that there are four to five hundred comets in the solar system. And if it is with comets as it is with the planets—if the largest are farthest from the Sun, and if the smallest are the only ones that approach closely enough for us to see them—what an immense volume of matter! What an enormous force on the body of this star; what a pressure, that is to say, what an internal friction on all parts of its mass, and by consequence what heat, and what fire, is produced by this friction!

Because, in our hypothesis, the Sun was a mass of molten matter, even before the expulsion of the planets. As a consequence, this fire then did not have a cause other than the pressure of this great number of comets that circulated before and circulate still today around this common center. If the original mass of the Sun had been diminished by one six-hundred-fiftieth by the expulsion of the material of the planets, since their formation, the total quantity of the source of its fire, that is to say, of the total pressure, was augmented in the proportion of the total pressure of the planets, united with the original pressure of all of the comets, with the exception of those that caused the expulsion, and of which the matter was mixed with those of the planets leaving the Sun: which, by consequence of this loss, only became more brilliant, more active, and more effective at lighting, heating, and making fecund its universe.

In pushing these deductions further, one will easily persuade oneself that the satellites that circulate around their principal planet, and that weigh upon them like the planets weigh upon the Sun—that the satellites, I say, must communicate a certain degree of heat to the planet around which they circulate. The pressure and the movement of the Moon must give the Earth a certain degree of heat, which would be larger if the speed of circulation of the Moon was greater. Jupiter, which has four satellites, and Saturn, which has five, with a great ring, must for this reason be animated with a certain degree of heat. If these planets, very far from the Sun, were not provided like the Earth with internal heat, they would be more than frozen; and the extreme cold that Jupiter and Saturn would have to support, due to their distance from the Sun, could be tempered only by the action of their satellites. The more numerous, greater, and rapid are the circulating bodies, the more the body that serves them as an axle or a pivot will be heated by the internal friction that they will impose on all parts of its mass.

These ideas connect perfectly with those that serve as a basis for my hypothesis on the formation of the planets; they are simple and natural consequences. But, I have the proof that few people have grasped the relations within, and the whole of, this great system. Nevertheless, is there a subject more exalted, more worthy of exercising the power of genius? People have criticized me without listening to me. How can I respond? If not to say that everything speaks to watchful eyes; everything is evident to those who know how to look; but nothing is visible, nothing is clear to the vulgar, and even to the vulgar savant made blind by prejudice. Let us try nevertheless to render the truth more palpable; let us increase the number of probabilities; let us make the likelihood greater; let us add light to light, in unifying the facts, in accumulating the proofs, and then let us judge without anxiety and without plea; because I have always thought that a writer should occupy himself solely with his subject, and not at all with himself, and it is contrary to good manners to want to occupy others with this. Consequently, personal criticisms must remain without response.

I admit that the ideas of this system can seem hypothetical, strange, and even chimerical to all those, judging things by the evidence of their senses only, who have never conceived how one knows that the Earth is only a small planet, expanded around the equator and depressed beneath the poles; to those who do not know how one is certain that the celestial bodies weigh upon each other, mutually acting and reacting—how one could measure their size, their distance, their movement, their weight, and so on. But, I can be persuaded how these same ideas will appear simple, natural, and even grand, to the small number of those who, through observations and the reflections stemming from them, are led to know the laws of the universe, and who judge things by their proper lights, and see them without prejudice, such as they are or they could be. Because, these two points of view are about the same. Those who see a clock for the first time would say that the principle of all of its movement stems from a spring, although it is in fact a weight. This would be misleading to the vulgar person only, while to the eyes of a philosopher it would explain the machine.

Thus it is that I do not affirm nor even positively contend that our Earth and the planets were formed necessarily and really by the shock of a comet that projected out of the Sun one six-hundred-fiftieth of its mass. But that which I would like to make understood, and that which I still uphold as a very probable hypothesis, is that a comet which, at its perihelion, would approach the Sun closely enough to furrow and groove its surface, could produce such an effect, and it is not impossible that new planets could be formed some day in a similar fashion, which all would circulate together like the present planets, in the same direction, and almost in the same plane around the Sun. Such planets would also spin, and with the matter having come from the Sun in a state of liquefaction, would be subject to centrifugal forces, and be raised at the equator and depressed at the poles; such planets could similarly have satellites in greater or lesser number, circulating around them along their equatorial planes, with motions being comparable to those of the satellites of our planets: such that all the phenomena of these possible and ideal planets would be (I do not say the same) but in the same order, and in similar relations to those of the phenomena of the present-day planets. And for proof, I ask only that one consider whether the motion of all the planets, in the same direction and virtually on the same plane, does not suggest a common impulsion? I ask if there are in the universe some bodies, excepting the comets, that could communicate such an act of impulsion? I ask if it is not probable that comets fall from time to time into the Sun, since that of 1680 has, so to speak, grazed its surface; and if consequently such a comet, on grooving this surface of the Sun, would not communicate its motion of impulsion to a certain quantity of matter that it would separate from the body of the Sun, projecting it beyond? I ask whether, in this torrent of projected matter, there would not be formed globes by the mutual attraction of the parts, and if these globes would not find themselves at different distances, following the different densities of the matter, and if the lighter ones would not be flung further than the denser ones by the same impulsion? I ask if the situation of all these globes, almost in the same plane, does not suggest clearly enough that the projected torrent was of considerable size, and was produced by a single impulsion, since all of the component matter is only slightly removed from a common direction? I ask how and where the matter of the Earth and the planets could have liquefied, if it had not resided in the very body of the Sun; and if one could find a cause for that heat and that glow of the Sun, other than that of its mass and of the internal friction produced by the action of all the enormous bodies that circulate around it? Finally I ask that one examine all of the evidence and that one follow all of the opinions and that one compare all of the analogies upon which I have founded my reasoning: one would be content to conclude with me that, if God had allowed it, it could be solely through the laws of Nature that the Earth and the planets had been formed by this same manner.

Let us thus follow our object, and from this time that has preceded time and that is withheld from our observations, let us pass to the first age of our universe, where the Earth and the planets having obtained their form, gained their consistency, and from liquid became solid. This change of state happened naturally and by the sole effect of the diminution of heat: the matter that makes up the earthly globe and the other planetary globes was in a state of fusion while they commenced to rotate. They therefore obeyed, like all other fluid matter, the laws of centrifugal force: those parts near the equator, which were subject to the greatest movement in rotation, became the highest; those that were near the poles, where this movement was lesser or absent, were depressed in the exact and precise proportion that is demanded by the laws of gravity when combined with those of centrifugal force,[2] and this form of the Earth and the planets has been conserved to this day, and will be conserved perpetually. Even if one would wish to suppose that the moment of rotation came to accelerate, because the matter, having passed from a state of fluidity to that of solidity, the cohesion of the parts alone suffices to maintain the primordial form, and to change it would need the speed of rotation to take on an almost infinite speed, that is to say, large enough that centrifugal force becomes greater than the force of cohesion.

Now the cooling of the Earth and the planets, like that of all hot bodies, started at the surface. The molten matter became consolidated in a fairly short time; as soon as the great heat that had penetrated them escaped, the parts of the matter that it had kept separate came closer together by their mutual attraction. Those that had sufficient rigidity to resist the violence of the fire formed solid masses; but those that, like air or water, were rarefied or volatized by the fire, could not make part of the same body as did the others, and separated from them in the first times of cooling. All the elements could be transmuted and converted, and the instant of the consolidation of fixed matter was also that of the greatest conversion of the elements and the production of volatile matter. These were reduced to vapor and far dispersed, forming around the planets a kind of atmosphere similar to that of the Sun; because one knows that the body of this star of fire is surrounded by a sphere of vapors, which extends for immense distances, and maybe as far as the Earth's orbit. The real existence of this solar atmosphere is demonstrated by a phenomenon that accompanies total eclipses of the Sun. The Moon to our eyes can cover its disc entirely, and nevertheless one can still see a limb or great circle of vapors, of which the light is sufficiently bright to illuminate us by about as much as does that of the Moon: without that, the terrestrial globe would be plunged into the most profound obscurity during the time of total eclipse. One has seen that this solar atmosphere is more dense in the regions neighboring the Sun, and that it becomes more rarefied and more transparent as it extends and distances itself farther from the body of this star of fire. One can thus not doubt that the Sun is surrounded by a sphere of aqueous material, aerial and volatile, which its violent heat keeps suspended and kept at immense distances, and that at the moment of the projection of the planets, the torrent of fixed material departing from the body of the Sun did, in traversing its atmosphere, entrain a great quantity of these volatile substances of which it is composed. These are the same volatile materials, aqueous and aerial, which then formed the atmosphere of the planets, and those were similar to the atmosphere of the Sun, as long as the planets were, like it, in a state of fusion or great incandescence.

All the planets were thus only masses of liquid glass, surrounded by a sphere of vapors. As long as this state of fusion endured, and even for a long time afterward, the planets were in themselves luminous, as are all incandescent bodies. But, as the planets gained solidity they lost their light. They became altogether dark only after they became consolidated to their center, and long after the consolidation of their surface, as one sees, in a mass of melting iron, the light and the redness persist for a very long time after the consolidation of the surface. In these first times, when the planets shone with their own fire, they must have sent out rays, thrown out sparks, made explosions, and then suffered, upon cooling, various ebullitions, as the water, the air, and other materials that could not support the fire fell to the surface. The production of the elements, and then their action and interaction, could not fail to produce irregularities, asperities, higher areas, and caverns at the surface and in the first-formed layers of the interiors of these great masses. It is to this epoch that one has to ascribe the formation of the highest mountains of the Earth, of the Moon, and all the asperities and irregularities that one sees on the planets.

Let us represent the state and aspect of our universe in its first age: all the planets, newly consolidated at the surface, were still liquid in their interiors, and threw out a vivid light; they were so many little suns detached from the great one, and which only ceded to it in volume, and from which light and heat spread out similarly. This time of incandescence lasted until the planet became consolidated to its center, that is to say, 2,936 years for the Earth, 644 years for the Moon, 2,127 years for Mercury, 1,130 years for Mars, 3,596 years for Venus, 5,140 years for Saturn, and 9,433 years for Jupiter.

The satellites of these two huge planets, just like the ring that surrounds Saturn, are all in the plane of the equator of their principal planet, having been projected out in the time of liquefaction by the centrifugal force of these huge planets that rotate with prodigious speed: the Earth of which the speed of rotation is about 9,000 leagues for twenty-four hours, that is to say, about six-and-a-quarter leagues per minute, at this time projected out of itself the least dense parts around its equator, which reassembled themselves by their mutual attraction at a distance of 85,000 leagues, where they formed the globe of the Moon. I propose here nothing that is not confirmed by the facts, when I say that it is the least dense parts that were projected out, and that they were in the region of the equator; because one knows that the density of the Moon is to that of the Earth as 702 is to 1,000, that is to say, more than a third less; and one knows also that the Moon circulates around the Earth in a plane which is not farther than twenty-three degrees from our equator, at which its average distance is around 85,000 leagues.

On Jupiter, which makes a rotation in ten hours, and of which the circumference is eleven times greater than that of the Earth, the speed of rotation is 165 leagues each minute: this enormous centrifugal force projected out a great torrent of material of different degrees of density, in which formed the four satellites of this huge planet, of which one, as small as our Moon, is only 89,500 leagues away from it, that is to say, it neighbors Jupiter much like the Moon does the Earth. The second, of which the matter is a little less dense than that of the first, and which is about the same size as Mercury, formed 141,800 leagues away; the third, composed of parts yet less dense, and which is about the same size as Mars, formed at 225,800 leagues; and finally the fourth, of which the matter is the lightest of all, was projected yet farther, and was only reassembled at 397,877 leagues, and all four are found more or less in the plane of the equator of their principal planet, and circulate in the same sense around it. Besides, the matter that makes up the globe of Jupiter is itself much less dense than that of the Earth. The planets neighboring the Sun are the most dense; those that are the most distant are in themselves the lightest; the density of the Earth is to that of Jupiter as 1,000 is to 292; and one can presume that the matter that forms the satellites is yet less dense than that of which it is itself composed.

Saturn, which probably turns around itself yet more quickly than Jupiter, not only produced five satellites but also a ring which, following my hypothesis, must be parallel to its equator, and which surrounds it like a suspended and continuous bridge at a distance of 54,000 leagues: this ring, greatly larger than thick, is made of solid opaque matter, and similar to that of the satellites; it was in the same state of fusion, and then incandescence. All of these great bodies have conserved their primitive heat in accordance with their thickness and their density. Hence, the ring of Saturn, which seems to be the thinnest of all the celestial bodies, is that which would have first lost its own heat, if it had not drawn very large supplements of heat from Saturn itself, which it is very close to. And then the Moon and the first satellites of Saturn and of Jupiter, which are the smallest of the planetary globes, would have lost their own heat, over times always proportional to their diameter; after which, the largest satellites would have lost their heat, and all would be colder than the globe of the Earth, if several of them had not received from their main planet an immense heat at the beginning. Finally, the two great planets, Saturn and Jupiter, still retain now a very large part of their heat by comparison with that of their satellites, and even with that of the globe of the Earth. Mars, of which the duration of rotation is twenty-four hours and forty minutes, and of which the circumference is only thirteen twenty-fifths that of the Earth, turns once more slowly than the terrestrial globe, its speed of rotation being barely more than three leagues a minute; by consequence its centrifugal force has always been less than half that of the terrestrial globe. It is for this reason that Mars, although less dense than the Earth in the proportion of 730 to 1,000, has no satellites at all. Mercury, of which the density is to that of the Earth as 2,040 is to 1,000, could have produced a satellite only by a centrifugal force more than double that of the globe of the Earth. But, although the speed of its rotation could not be observed by the astronomers, it is more than likely that rather than being twice that of the Earth, it is much less. Thus one can believe with justification that Mercury has no satellites.

Venus could have one, because being a little smaller than the Earth in the proportion of seventeen to eighteen and turning a little more quickly in the relation of twenty-three hours and twenty minutes to twenty-three hours and fifty-six minutes, its speed is more than six-and-three-quarter leagues a minute, and by consequence its centrifugal force is around one-thirteenth greater than that of the Earth. This planet could thus have produced one or two satellites at the time of its liquefaction, if its density, greater than that of the Earth in the proportion of 1,270 to 1,000, that is to say, by more than five to four, was not opposed to the separation and the projection of even its most liquid parts. It could be for this reason that Venus did not have any satellites, although there are some observers who claim to have glimpsed one around this planet.

To all these facts that I have just expounded, one can add another, which has just been communicated to me by M. Bailly, the learned physicist-astronomer of the Académie des Sciences. The surface of Jupiter is, as one knows, the subject of distinct changes, which seem to indicate that this great planet is still in a state of fluidity and boiling. Taking thus, in my system of general incandescence and cooling of the planets, the two extremes, that is to say Jupiter as the largest, and the Moon as the smallest of all the planetary bodies, it is found that the former, which has not yet had the time to chill and to become entirely solid, shows us at its surface the effects of an internal movement that is provoked by fire; while the Moon which, by its smallness, could cool in a matter of centuries, shows us only a perfect calm, that is to say, a surface that is always the same and on which one sees neither movement nor change. These two facts known to astronomers link with other analogies that I have presented on the subject, and add a small degree more of probability to my hypothesis.

By the comparison that we have made of the heat of the planets to that of the Earth, one has seen that the time of incandescence of the terrestrial globe has lasted two thousand, nine hundred and thirty-six years; and that of its heat, to the point of not being able to touch it, has been thirty-four thousand, two hundred and seventy years, which makes in total thirty-seven thousand, two hundred and six years; and that is the first moment of the possible birth of a living Nature. Up until then the elements of the air and the water were still conjoined, and could not separate nor rest on the burning surface of an Earth that would dissipate them into vapor; but once this glow had cooled, a benign and fecund heat succeeded by degrees the devouring fire that had opposed all production, and even the separation, of the elements. The element of fire, in those first times, had, so to say, absorbed the others. None existed separately: earth, air, and water conjoined together with fire, offered, instead of their distinct forms, only a burning mass of inflamed vapors. It was thus only after thirty-seven thousand years that the people of the Earth could date the events of the world, and calculate the facts of an organized Nature.

One needs to refer, to this first epoch, that which I have written on the state of the heavens in my Memoirs on the temperature of the planets. All, at the beginning, were brilliant and luminous. Each was a little sun, of which the heat and light diminished little by little and successively dissipated with the passage of time, as I have already said, following my experiments on the cooling of bodies in general, of which the duration is always very close to being proportional to their diameters and to their density.

The planets, like their satellites, thus cooled, some earlier and some later; and, in losing part of their own heat, they lost all of their own luminosity. The Sun alone is maintained in its splendor because it is the only one around which circulate a sufficiently great number of bodies to maintain its light, heat, and fire.

But without insisting for any longer on these objects, which appear so far from our view, let us gaze upon the sole globe of the Earth. Let us pass to the second epoch, that is to say, to the time when the material that composes it, on becoming consolidated, formed great masses of vitrescible material.

I must respond to only one type of objection that has already been made to me, on the very long duration of time. Why throw us, as one has said to me, into an interval as vague as a duration of one hundred and sixty-eight thousand years? Because in your portrayal, the Earth is seventy-five thousand years old, and a living Nature must yet continue for ninety-three thousand years. Is it easy, is it even possible to form an idea of all or of part of such a long succession of centuries? I have no reply other than the exposition of the monuments and the consideration of the works of Nature. I will give you the detail and the dates in the epochs that will succeed this one, and one will see that far from having to unnecessarily augment the duration of time, I have perhaps shortened it too greatly.

And why does the human spirit seem to lose itself in an extent of time, rather than one of space or in a consideration of measures, weights, numbers? Why are a hundred thousand years more difficult to conceive and to count than a hundred thousand pounds in money? Is it because a summation of time cannot be touched or made visible? Or is it not rather that, in being accustomed to our all-too-short existence, we regard a hundred years as a great amount of time, and we have difficulty in forming an idea of a thousand years, and cannot any more imagine ten thousand years, nor conceive of a hundred thousand? The only means is to divide these long periods of time into several parts, to compare in spirit the duration of each of these parts with their great effects, and above all with the constructions of Nature; to make insights into the number of centuries that were needed to produce all of the animals and shells with which the Earth is replete; and then, on the yet greater number of centuries which have passed for the transport and deposition of these shells and their detritus; and finally, upon the number of other centuries that followed, that were necessary for the petrifaction and the desiccation of this matter. And from that, one will feel that this enormous duration of seventy-five thousand years, that I have counted from the formation of the Earth up to its present state, is not yet long enough for all of the great works of Nature, the construction of which shows us that which could be made only by a slow succession of regulated and constant changes. To render this insight more obvious, let us give an example. Let us seek to know how much time it would take to make a hill in shale a thousand toises in height: the successive water-formed sediments made up all of the beds of which the hill is composed, from the bottom to the summit. We can judge the successive and daily deposition from the waters by the layers of the shale: these are so thin that one can count a dozen in a line of thickness. Let us suppose thus that each tide deposits sediment that is a twelfth of a line in thickness, that is to say, of one-sixth of a line each day, the deposit will augment by a line in six days, by six lines in thirty-six days, and by consequence by about five inches in one year. That gives more than fourteen thousand years for the time necessary for the construction of one hill of clay one thousand toises in height. Even this time will seem too short, if one can compare it with what passes before our eyes on certain seashores, where silts and muds are deposited, such as on the Normandy coast;[3] because the deposit augments only insensibly and by much less than five inches each year. And, if this hill of shale is crowned with limestone rocks, the duration of time, which I reduced to fourteen thousand years, has to be augmented, does it not, by that which was necessary to transport the shells with which the hill is capped? And, was not this very long time followed by the time necessary for the petrifaction and the desiccation of these sediments, and again by a time long enough for the shaping of the hill into salient and re-entrant angles? I have thought necessary to enter into this detail beforehand, so as to demonstrate that, instead of pushing too far the limits of time, I have constrained them as far as was possible for me, without evidently contradicting the facts consigned in the archives of Nature.

SECOND EPOCH

When Matter, Being Consolidated, Formed the Interior Rock of the Globe and the Great Vitrescible Masses That Are at Its Surface

---

We have seen that, in our hypothesis, two thousand, nine hundred and thirty-six years must have elapsed before the terrestrial globe could have taken on its consistency and that its entire mass was consolidated as far as its center. Let us compare the effects of this solidification of the fused globe of the Earth to that which we see happening on a mass of molten metal or glass, while it begins to cool: holes, undulations, and asperities form on the surface of these masses; and below the surface there are spaces, cavities, blisters, which can represent here the first irregularities that were present on the surface of the Earth and the cavities in its interior. We will have from that an idea of the great number of mountains, of valleys, of caverns, and crevices that formed from these times in the external layers of the Earth. Our comparison is all the more exact, in that the highest mountains, which I presume to be three thousand or three thousand, five hundred toises in height, are, by comparison with the diameter of the Earth, in relation only an eighth of a line on a globe two feet in diameter. Just as the mountain chains seem to us so prodigious, as much by their volume as by their height, they are in reality only small irregularities, proportionally to the size of the globe, and which could not help but form once it attained its solidity; these are natural effects produced by an altogether natural and very simple cause, that is to say, by the action of cooling on matter in fusion, while it consolidates at the surface.

It was then that the elements were formed, by the cooling and during its progress. Because, at this epoch, and even for long afterward, as long as the excessive heat lasted, there was a separation and even a projection of all the volatile parts, such as the water, the air, and the other substances that the great heat drove out, and which could exist only in a region that was then more temperate than was the surface of the Earth. All of these volatile materials thus extended around the globe in the form of an atmosphere at a great distance where the heat was less strong, while the fixed materials, melted and glassy, were being consolidated. They formed the interior rocks of the globe and the kernels of the great mountains, of which the summits, the interior masses, and the bases are in effect composed of vitrescible matter. Thus the first local establishment of the great mountain chains belong to this second epoch, which preceded by many centuries that of the formation of the limestone mountains, which existed only after the first appearance of the waters, because their composition suggest the production of shells and of other substances that the sea fomented and nourished. As long as the surface of the globe had not chilled to the point of allowing water to rest there without being exhaled as vapor, all of our oceans were in the atmosphere. They could fall and settle on the Earth only at the moment when its surface had cooled enough to no longer boil off the water. And, this time of the settling of waters on the surface of the globe only preceded by a few centuries the moment when one could touch this surface without being burnt. From which, in counting seventy-five thousand years since the formation of the Earth, and half of this time for its cooling sufficiently to touch, there passed perhaps twenty-five thousand of these first years before water, from always being rejected into the atmosphere, could establish a presence on the surface of the globe. Because, although there is a large enough difference between the degree at which hot water stops hurting us and that at which it comes to boil, and there is yet a considerable distance between this first stage of boiling and that at which water is immediately dispersed as vapor, one can nevertheless be assured that this difference in temperature could not be greater than I allow here.

Thus, in these first twenty-five thousand years the terrestrial globe, at first luminous and hot like the Sun, lost its light and fire only little by little; its state of incandescence lasted two thousand, nine hundred and thirty-six years, because this time was required for it to become consolidated to its center. Then, the fixed materials of which it was composed became even more fixed in contracting more and more by the cooling. Little by little, they took on the nature and consistency that we recognize today in the rocks of the globe and of the high mountains, which are in effect composed, in their interior and to their summit, only by materials of the same nature[1]—and so their origin dates from this same epoch.

It is also, in these first thirty-seven thousand years, that were formed, by sublimation, all of the great veins and the thick lodes in the mines where metals are found. The metallic substances were separated from the other vitrescible materials by the long and constant heat, which sublimated them and drove them into the interior of the mass of the globe in all of the projections at the surface, where the contraction of these materials, caused by more rapid cooling, left cracks and cavities, which were encrusted and at times filled with these metallic substances that we find today.[2] Because it is necessary, with regard to the origin of the ores, to make the same distinction that we have indicated for the origin of vitrescible matter and calcareous matter, of which the first were produced by the action of fire and the others through the intermediary of water. In metal ores, the principal lodes or, if one wishes, the primordial masses, were produced by fusion and by sublimation, that is to say, by the action of fire; and the other ores, which one can regard as secondary and parasitic lodes, were only produced later, by the action of water. These principal lodes, which seem akin to the trunks of metallic trees, having all been formed either by fusion in the time of primitive fire, or by sublimation in subsequent times, were found and are still found today as perpendicular fissures in the high mountains; while at the foot of these same mountains there are the small veins that one can at first take to be the branches of these metallic trees, but of which the origin is nevertheless very different. Because these secondary ores were not formed by fire, but were produced by the repeated action of water which, in these later times, detached particulate minerals from these ancient lodes, which it carried and deposited in different forms, and always below the primitive lodes.[3]

The production of these secondary ores being thus much more recent than that of the primordial ores, and supposing the action and mediation of water, the formation must, like that of the calcareous matters, be linked with subsequent epochs, that is to say, with the time when the burning heat having cooled, the temperature of the surface of the Earth allowed waters to settle—and then, at the time when these same waters, having exposed our continents, the vapors began to condense on the mountains, to produce sources of running water there. But before this second and this third time, there were other great effects, which we have to demonstrate.

Let us picture to ourselves, if that is possible, the aspect that the Earth offered at this second epoch, that is to say, immediately after its surface had gained its solidity, and before the great heat allowed water to rest there or even to fall from the atmosphere. The plains, the mountains, just as the interior of the globe, were equally and uniquely composed of matter melted by fire, all made glassy, all of the same nature. If one imagines for a moment the surface of the globe then bereft of all its oceans, of all of its limestone hills, as well as all of its horizontal beds of stone, of chalk, of tufa, of vegetable earth, of shale, in a word of all the liquid and solid materials that were formed or deposited by the waters—what would that surface be like after the lifting off of these immense wastes? There would remain only the skeleton of the Earth, that is to say, the vitrescible rock that makes up its interior mass. There would remain the perpendicular fissures produced in the time of consolidation, augmented and enlarged by the cooling; there would remain the metals and fixed minerals which, separated from the vitrescible rock by the action of fire, filled by fusion or by sublimation the perpendicular fissures of these prolongations of the interior rock of the globe. And finally, there would remain the holes, the crevices, and all of the internal cavities of this rock that is the basis of, and that serves as the support for, all of the terrestrial matter later brought in by the waters.

And just as the fissures produced by cooling cut and cross the vertical plan of the mountains, not only from top to bottom, but from the front, from the back, and from one side to the other, in each mountain they took the general direction of its initial form. The result is that ores, especially those of precious metals, should be searched for using a compass, always following the direction found in the discovery of the first lode—because in each mountain, the perpendicular fissures that traverse it are nearly parallel. Nevertheless, one should not conclude, as certain mineralogists have done, that one should always seek metals along the same direction—for example, along the line of eleven o'clock or that of noon. Because, often an ore along noon or eleven o'clock can be cut by a lode at eight or nine o'clock, and so on, which extends branches in different directions; and besides, one sees that, following the different form of each mountain, the vertical fissures that traverse it are truly parallel to each other, but their direction, although shared within the same place, has nothing in common with the direction of the perpendicular fissures of another mountain, at least if that second mountain is not aligned parallel to the first.

The metals and most of the metallic minerals are thus the work of fire, since one finds them only within the fissures of vitrescible rock, and in these primordial ores, one never sees either shells or any other debris from the sea mixed with them. The secondary ores, which are found, by contrast and in small quantity, in the calcareous rock, in schists or shales, were formed later at the expense of the first ones, and by the action of water. The flakes of gold and of silver carried by a few rivers certainly came from these first metallic lodes, enclosed within the high mountains. Metal particles that are yet smaller and more fragile can, in being concentrated, form small new ores of the same metals. But, these parasitic ores, which take a thousand different forms, belong, as I have said, to times that are very modern by comparison with those of the first lodes, which were produced by the action of primitive fire. Gold and silver, which can long remain in a state of fusion without being noticeably altered, often show themselves in their native form. All the other metals commonly show themselves only in mineralized form, because they were formed later, by the combination of air and water that entered into their composition. Besides, all the metals are susceptible to being volatilized by fire at different degrees of heat, by which they become successively sublimated during the progress of cooling.

One could think that if there are fewer gold and silver mines in northern countries than in the countries of the south, it is because there are commonly only small mountains in the lands of the north by comparison with those of the southern countries. The primitive matter, that is to say, vitreous rock, in which alone gold and silver are formed, is much more abundant, much more elevated, and much better exposed in the countries of the south. These precious metals seem to be the immediate product of fire. The gangues and other matter that accompany them in their ore are themselves of vitrescible materials; and as the veins of these metals formed, either by fusion or by sublimation, in the first times of cooling, they are found in the greatest quantities in the high mountains of the south. The less perfect metals, such as iron and copper, which are less fixed by fire, because they contain some materials that fire can volatilize more easily, were formed in later times; thus one finds them in substantially greater quantity in the countries of the north than those of the south. It even seems that Nature had assigned different metals to different climates of the world: gold and silver to the hottest regions; iron and copper to the coldest countries, and lead and tin in temperate countries. Similarly, it seems that Nature placed gold and silver in the highest mountains; iron and copper in moderate-sized mountains; and lead and tin in the lowest. It seems also that, while primordial ores of the different metals are found in vitrescible rock, those of gold and silver are sometimes mixed with other metals: that iron and copper are often accompanied by matter that suggests water as an intermediary, which seems to prove that they were not formed at the same time; and with regard to tin, lead and mercury, there are differences that seem to indicate that they were produced in very different times. Lead is the most vitrescible of all of the metals, and tin is the least. Mercury is the most volatile of all, and yet it differs from gold, which is the most fixed of all, only by the degrees of fire needed for their sublimation, because gold, like all the other metals, can equally be volatilized by a more or less great heat. Thus all of the metals were sublimated and volatilized successively, as cooling progressed. And as only a very gentle heat is needed to volatilize mercury, and a moderate heat suffices to melt tin and lead, these two metals remained liquid and flowing for much longer than the four first metals. Mercury remains liquid, because the present temperature of the Earth is more than sufficient to keep it in fusion. It will become solid only when the globe is cooled by a fifth more than it is today, because it needs to be 197 degrees below the present temperature of the Earth for the fluid metal to solidify, which is about one-fifth part of the 1,000 degrees below that of the Earth's solidification.

Lead, tin, and mercury thus percolated successively, by their fluidity, into the lowest parts of the rocks of the globe, and they were, like the other metals, sublimated in the fractures of the high mountains. The ferruginous materials that could support a very violent heat, without melting enough to flow, formed in the countries of the north, in metallic accumulations so considerable that one finds there mountains entirely of iron,[4] that is to say, of a vitrescible and ferruginous rock, which often yields seventy pounds of iron per quintal. Those are the ores of primitive iron; they occupy very great areas of the countries of our north; and their substance being only iron produced by fire, these ores remain susceptible to magnetic attraction; as is all ferruginous matter that has been subject to fire. Lodestone is of the same nature; it is only a ferruginous stone, of which there are great masses and even mountains in some countries, and especially in those of our north.[5] It is for this reason that a magnetic needle always points toward those countries where all of the iron mines are magnetic. Magnetism is a constant effect of constant electricity, produced by internal heat and by the rotation of the globe. But, if it depended solely on this general cause, the magnetic needle would point always and everywhere in the direction of the pole. Now, the different declinations seen in the different countries, even under the same parallel, show that the particular magnetism of mountains of iron and lodestone considerably influence the direction of the needle, since it deviates more or less to the right or left of the pole, according to where it is and according to the greater or lesser distance of the mountains of iron.

But let us return to our principal object, to the topography of the globe, before the falling of the waters. We have only a few traces still remaining of the first form of its surface. The highest mountains composed of vitrescible matter are the sole witnesses of this ancient state; they were then yet more elevated than they are today. Because, since this time and after the establishment of the waters, the movements of the sea, and then the rains, the winds, the frosts, the air currents, the fall of torrents, and all of this damage from the elements of the air and of the water, and the shaking from underground movements, have not ceased to degrade them, to cut them, and even to topple the least solid parts, and we cannot doubt that the valleys that are at the foot of the mountains were much deeper than they are today.

Let us try to provide a glimpse, or rather an enumeration of these primitive heights of the globe. (1) The chain of the Cordillera, or the mountains of America, extends from the point of Tierra del Fuego to the north of New Mexico and finally ends in those northern regions that are not yet known. One can look at this chain of mountains as continuing along a length of more than 120 degrees, that is to say, of three thousand leagues, because the strait of Magellan is only an accidental cut, and subsequent to the local establishment of this chain, of which the highest summits are in the country of Peru, with a lowering almost equally to the north and to the south. It is thus under the very equator that are found the most elevated parts of this primitive chain of the highest mountains in the world; and we will see this as remarkable, that from this point at the equator they fall in height almost equally to the north and to the south, and also that they end at about the same distance, that is to say at fifteen hundred leagues on each side of the equator. In this way, there remains at each extremity of this range of mountains only around thirty degrees, that is to say, seven hundred and fifty leagues of sea or of unknown land to the southern pole, and an equal distance of which a few coasts are known toward the north pole. This chain is not under exactly the same meridian, and does not form a straight line. It curves first to the east, from Valdivia* as far as Lima, and its greatest deviation is found under the Tropic of Capricorn. Then, it advances to the west, returns toward the east near Popayan, and from there it curves strongly to the west, from Panama up to Mexico; after which it returns to the east, from Mexico up to its extremity, which is at thirty degrees from the pole, and which ends nearly at the islands discovered by de Fonté. In considering the situation of this long run of mountains, one must again see as a very remarkable thing that they are all much closer to the seas of the west than to those of the east. (2) The mountains of Africa, of which the principal chain is called by several authors *The Spine of the World*, is also strongly elevated, and extends from the south to the north, like that of the Cordillera in America. This chain, which forms in effect the spine of the back of Africa, begins at the Cape of Good Hope and runs almost under the same meridian to the Mediterranean Sea, by the point of the Morea.† We can also see, as a very remarkable thing, that the middle of this great chain of mountains, about fifteen hundred leagues long, is found exactly under the equator, like the midpoint of the Cordillera. From this one can hardly doubt that the most elevated parts of the great chains of mountains in Africa and in America are both found beneath the equator.

In these two parts of the world, of which the equator traverses fairly exactly the continents, the principal mountains are thus aligned from the south to the north; but they throw very considerable branches to the east and to the west. Africa is traversed from the east to the west by a long run of mountains, from Cape Gardafui as far as the islands of the Cape Verde. The Atlas Mountain also cuts it from the east to the west. In America, a first branch of the Cordillera traverses the Magellanic islands from the east to the west; another extends in the same direction to Paraguay and in all of the extent of Brazil; a few other

* N. b., in the original this is Baldivia, likely a misprint. Valdivia is named after Pedro de Valdivia.

† Now the Peloponnese.

branches extend from Popayan in the mainland as far as Guyana. Finally, if we keep following this great chain of mountains, it will seem to us that the Yucatan Peninsula, the islands of Cuba, of Jamaica, of Saint-Domingue, Puerto Rico, and all of the Antilles, are only a branch that extends from the south to the north, from Cuba and the end of Florida up to the lakes of Canada, and from there runs from the east to the west to rejoin the extremity of the Cordillera, beyond the Sioux lakes. (3) In the great continent of Europe and Asia, which not only is not, like those of America and of Africa, traversed by the equator, but is even far from it, the chains of the principal mountains, rather than being directed from the south to the north, are from the west to the east: the longest of these chains begins at the bottom of Spain, reaches the Pyrenees, extends in France through the Auvergne and the Vivarais,* passes then via the Alps, in Germany, in Greece, in Crimea, and reaches the Caucasus, the Taurus, the Imaus, which neighbors Persia, Kashmir, and the Mughal to the north, up to Tibet, where it extends into the Chinese Tartary, and finishes against the land of Yeco. The principal branches thrown off this chain are directed from the north to the south in Arabia, as far as the straits of the Red Sea; in Hindustan, up to Cape Comorin; from Tibet, up to the point of Malacca. These branches do not form suites of particular mountains whose summits are very high. On the other side, this principal chain throws a few branches from the south to the north, which extend from the Alps of the Tyrol as far as Poland; then from Mount Caucasus to Moscow, and from Kashmir as far as Siberia; and these branches that go from the south to the north of the principal chain do not show mountains that are as high as those of the branches of this very same chain that extend from north to south.

Here is thus more or less the topography of the surface of the Earth in the times of our second epoch, immediately after the consolidation of the matter. The high mountains that we have just described are the primitive heights, that is to say, the asperities produced at the surface of the globe at the moment at which it took on its solidity. They must owe their origin to the effect of fire, and are also for this reason composed, in their interior and up to their summits, of vitrescible matter. All have at their base the interior rock of the globe, which is of the same nature. Several other smaller elevations traversed the surface of the Earth at the same time almost all in different directions. One can also be assured that, in all of the places where one finds mountains of crystalline rock or of other solid and vitrescible matter, their original and their local emplacement can be attributed only to the action of fire and the effects

* Approximately equals the Ardèche region of France.

of solidification, which never happens without leaving irregularities on the surface of the whole mass of molten matter.

At the same time as these causes produced the prominences and depths at the surface of the Earth, they also formed blisters and cavities in the interior, above all in the most exterior layers. Thus the globe, from the time of this second epoch, while it had gained its solidity and before the seas formed, showed a surface wrinkled with mountains and creased with valleys. However, all of the actions subsequent to and later than this epoch have competed to fill all of the exterior depths and even the interior cavities. These subsequent actions have also altered, almost everywhere, the form of these primitive irregularities. Those that were elevated only to a moderate height were for the most part later covered by sediments laid down from water, and all were surrounded at their bases up to great heights by the same sediments. It is for this reason that we have no other apparent witnesses of the first form of the surface of the Earth other than the mountains composed of vitrescible matter that we have listed. Nevertheless, these witnesses are sure and sufficient; because the highest summits of these first mountains were never overtopped by the waters, or at least never for more than a short time, as one finds there no debris of marine production, and they are only made of vitrescible matter. One cannot doubt that they owe their origin to fire, and that these prominences, like the interior rock of the globe, make up together a continuous body of the same nature, that is to say, of vitrescible matter, of which the formation preceded that of all the other matter.

In dividing the globe at the equator and comparing the two hemispheres, one sees that one of our continents contains proportionally much more land than the others, because Asia alone is more extensive than the parts of America, of Africa, of New Holland, and all that has been discovered of the lands beyond. There were thus fewer prominences and asperities on the southern hemisphere than on the northern, even from the time of the solidification of the Earth. If one considers for a moment this general disposition of the lands and of the oceans, one will recognize that all the continents narrow to the south while all the oceans correspondingly enlarge to the south. The narrow point of South America, that of California, that of Greenland, the point of Africa, that of the two peninsulas of India, and finally that of New Holland, clearly show this narrowing of the lands and expansion of the oceans toward the southern regions. This seems to show that the surface of the globe originally had more profound valleys in the southern hemisphere, and a greater number of prominences in the northern hemisphere. We will soon draw a few conclusions from this general disposition of the continents and of the oceans.

The Earth, before it received its waters, was thus irregularly bristling with asperities, deep hollows, and irregularities similar to those that we see on a melted block of glass or metal. It similarly had blisters and internal cavities, of which the origin, like those of the external irregularities, may be attributed only to the effects of solidification. The greatest peaks and exterior hollows and internal cavities were then found and are still found today beneath the equator between the two tropics, because this zone of the surface of the globe was the last to be solidified. And, as in this zone the movement of rotation is most rapid, it would have produced the greatest effects. The molten matter there being elevated more than elsewhere, and being cooled the last, more irregularities would have formed than on those other parts of the globe where the movement of rotation was slower and cooling more prompt. Thus one finds in that zone the highest mountains, the most intricately shaped seas sprinkled with an infinite number of islands, at the sight of which one cannot doubt that from its origin this part of the Earth was the most irregular and the least solid of all.[6]

And, as the molten matter must have come equally from both poles to elevate the equator, it seems, on comparing the two hemispheres, that our pole is a little less well provided than the other, because there is much more land and less sea from the Tropic of Cancer to the north pole; and, by contrast, there is much more sea and much less land from the Tropic of Capricorn to the other pole. The deepest valleys were thus formed in the cold and temperate zones of the southern hemisphere, and most solid and elevated lands are found in those of the northern hemisphere.

The globe then was, as it still is today, expanded at the equator by a thickness of about six and a quarter leagues. But, the superficial layers of this thickness were there riddled with cavities, and cut at its exterior into prominences and depths greater than elsewhere. The rest of the globe was furrowed and traversed in different directions by asperities that always became smaller as they approached the poles. All were composed only from the same molten matter, of which the interior rock of the globe is also composed; all owe their origin to the action of primitive fire and general vitrification. And so the surface of the Earth, before the arrival of the waters, showed only those first asperities that still form today the cores of our highest mountains. Those that were less elevated, having been since covered by sediments brought by water and by the debris of the products of the sea, are not as clearly known to us as those first mountains. One often finds beds of limestone above granite rocks, of crystalline rock, and other masses of vitrescible material; but one does not see masses of solid bedrock above limestone layers. We can thus be assured, without fear of being mistaken, that the rock of the globe is continuous with all of the high and low peaks that are of the same material: that is to say, of vitrescible matter. These peaks are part of the mass of the solid Earth; they are only very small prolongations, of which the least elevated were later covered by glass scoria, sand, muds, and all the debris of the products of the sea, which were carried and deposited by water in subsequent times: they are the subject of our third epoch.

THIRD EPOCH

When the Waters Covered Our Continents

Thirty or thirty-five thousand years after the formation of the planets, the Earth had cooled enough to receive the waters without dispersing them into vapor. The chaos of the atmosphere had begun to become organized: not only the waters, but all the volatile materials that the excessive heat had kept dispersed and suspended, fell successively onto the ground. They filled all the depressions, covered all the plains and all the ground that lay between the elevations of the surface of the globe, and they even overtopped all those that were not excessively high. There is clear proof that the seas covered the continent of Europe up to fifteen hundred toises above the level of the present-day sea,[1] since shells and other products of the sea can be found in the Alps and in the Pyrenees up to this same height. One has the same proofs for the continents of Asia and Africa, and even for those of America, where the mountains are higher than in Europe: fossil shells have been found at more than two thousand toises in height above the surface of the southern sea. It is thus certain that in these first times, the diameter of the globe was two leagues greater, since it was enveloped in water up to two thousand toises in height. The surface of the Earth was in general much higher than it is today. For a long time the seas covered it entirely, with the possible exception of some very high ground and of high mountain summits that alone were raised above this universal sea, of which the elevation was at least at the height at which one no longer finds shells. From this, one must infer that the animals to which these remains belonged can be regarded as the first inhabitants of the globe, and that this population was countless, to judge by the immense quantity of their remains and their detritus. It is these same remains and their detritus that formed all of the beds of limestone rock, of marble, of chalk, and of tufa that compose our hills and that extend across wide regions of all parts of the Earth.

Now at the beginnings of this sojourn of waters on the surface of the globe, did they not possess a degree of warmth that our fish and shellfish of today could not tolerate? And are we not forced to presume that the first products of a sea that was still boiling were different from those that are present today? This great heat could be suitable only for other types of shells and fish. Consequently, it is in the first times of this epoch, that is to say, from thirty to forty thousand years from the formation of the Earth, that one can report on the existence of lost species of which no living analogues can be found anywhere. These first species, now annihilated, lived for the ten or fifteen thousand years that followed the times when the waters were established.

And one should not be at all astonished in what I suggest here, that the fish and other aquatic animals were able to support a degree of heat much greater than the present-day temperature of our tropical seas. Because even today, we know of species of fish and plants that live in waters that are almost boiling, or at least reaching fifty to sixty degrees[2] on the thermometer.

But so as not to lose the thread of the many and great phenomena that we have to describe, let us go back to those earlier times, when the waters—until then in the form of vapor—were condensed and began to fall on an Earth that was burning, dry, parched, crevassed by fire. Let us try to represent the prodigious effects that accompanied and followed this sudden downpour of volatile materials, all separated, combined, sublimated in the time of solidification and during the progress of the first cooling. The separation of the element of the air and the element of water, the shock of the winds and the spray that fell in gusts on a smoking Earth; the cleansing of the atmosphere, which previously the rays of the Sun could not penetrate; this same atmosphere newly obscured by clouds of thick fumes; the redistillation, a thousand times repeated, and the continual boiling of water alternately fallen and repulsed; finally, the cleansing of the air, by the loss of the previously sublimated volatile matters, which all separated and descended more or less rapidly. What motions, what tempests must have preceded, accompanied, and followed the local establishment of these elements! And do we not have to relate to these first moments of shock and agitation, the upheavals, the first degradations, the eruptions and changes that gave a second form to the greater part of the surface of the Earth? It is easy to sense that the waters, which covered then almost everything, being continually agitated by the speed of their fall, by the action of the Moon on the atmosphere and on the waters that had already fallen, by the violence of the winds, and so on, would have been subject to all these impulsions, and in their movement they would have begun to carve more deeply the valleys of the Earth, to topple the less solid peaks, lowering the mountain crests, piercing their chains at their weakest parts; and after they were established, these same waters would have opened subterranean passages, destabilized the roofs of caverns, made them collapse, and consequently these same waters were lowered successively through filling the new depressions that they themselves had just formed. The caverns were the work of fire; the waters, once arrived, began to attack them: it destroyed them, and continues to destroy them. We must thus attribute the lowering of the waters to the collapse of caverns, as the sole cause that is demonstrated to us by the facts.

Here are the first effects produced by the mass, by the weight, and by the volume of water; but it produced others by its sole quality alone: it seized all the matter that it could soak and dissolve; it combined with air, earth, and fire to form the acids, the salts, and so on; it converted the scoria and the powder of primitive glass into clays; then, by its movement, it transported from place to place these same scoriae and all of the matter that was reduced into small particles. There was thus made in this second period, from thirty-five to fifty thousand years ago, such a great change to the surface of the globe, that the universal sea, at first very high, successively lowered to fill the depressions formed by the collapse of caverns whose natural roofs were sapped or penetrated by the action and effect of this new element and could no longer sustain the weight of the earth and the waters that pressed on them. As they were extensively collapsed by the rupture of one or of many caverns, the surface of the Earth, being depressed in these places, the water arrived from all sides to fill these new depths, and hence the general level of the sea diminished in proportion. In this way, from being at first at two thousand toises in height, the sea was progressively lowered to the level where we see it today.

One must presume that the shells and other marine products that one finds at great heights above the present level of the sea are the most ancient species in Nature; and it seems important to natural history to collect a large enough number of these productions of the sea that are found at this great height, and to compare them with those that are found on lower ground. We are assured that the shells of which our hills are made belong in part to unknown species, that is to say, to those species that no sea frequented by humans offers to us living analogues. If one ever makes a collection of these petrifactions, taken from the greatest heights of the mountains, then one will perhaps be able to pronounce on the more or less great age of these species, relative to others. All we can say of this today is that some of these monuments, which show us the existence of certain terrestrial and marine animals of which we do not know living analogues, show us at the same time that these animals were much larger than any species of the same kind living today: these great molar teeth with rounded points, eleven or twelve pounds in weight; these horns of Ammon, seven or eight feet in diameter and a foot thick, of which are found the petrified molds, are certainly of gigantic beings among types of quadruped and of shells. Nature was then in its first force, and worked the living and organic matter with a more active power in a hotter temperature; this organic matter was more divided, less combined with other matters, and could unite and combine with itself in great masses to develop in greater dimensions; this cause is sufficient to give a reason for all these gigantic productions that seem to have inhabited these first ages of the world.[3]

In making the seas fertile, Nature also spread the principles of life on all the lands that the water had not been able to submerge or that it had quickly abandoned; and these lands, like the seas, could be inhabited only by animals and plants capable of enduring a heat greater than that which is suitable for a living Nature today. We have monuments drawn from the heart of the Earth, and particularly from the bottom of mines of coal and slate, that show us that some of the fish and plants these materials contain are not of species existing today.[4] One can thus believe that the animal population of the sea is not older than that of the plants on land: the monuments and the witnesses are more numerous, more evident for the sea; but those that speak for the land are equally certain, and seem to show us that the ancient species of marine animals and of terrestrial vegetation were destroyed, or rather ceased to multiply, once the land and the sea lost the great heat necessary for their propagation.

The shells and the plants of these first times having prodigiously multiplied during this long interval of twenty thousand years, the shelled animals, the coral polyps, the madrepores, the starfish, and all the small animals that converted the water of the sea into stone, as they perished, abandoned their remains and their works to the vagaries of the waters. These will have transported, broken, and deposited these remains in thousands and thousands of places, because it is at this same time that the movement of the tides and the regular winds started to form the horizontal beds of the surface of the Earth by the sediments and the deposits of the waters. Then, the current gave to all the hills and all the low mountains their corresponding angles; so that their salient angles are always opposed to their reentrant angles. We will not repeat here what we have already said on this subject in our *Theory of the Earth*, and we will content ourselves with saying that this general disposition of the Earth's surface in corresponding angles, just like its composition in beds that are horizontal, or equally and in parallel inclined, clearly shows that the structure and form of the present-day surface of the Earth has been shaped by the waters and formed by its sediments. There were only the crests and peaks of the highest mountains which, perhaps, found themselves out of reach of the waters, or were submerged for a short while only, and on which by consequence the sea did not leave any imprints; but, not able to attack their summits, it took hold of their bases. It covered or penetrated the lower parts of these primitive mountains. It surrounded them with new matter, or pierced the roofs that sustained them. Often, it tilted them. Finally, it transported into its internal cavities combustible matters derived from the breakdown of vegetation, and pyrite, bituminous, and mineral matters, pure or mixed with earth and with sediments of all types.

The production of clays seem to have preceded that of shells, because the first action of the water was to transform the scoria and glassy powder into clays: hence the beds of clay formed some time before the banks of limestone rock; and one sees that these deposits of clayey matter preceded those of calcareous matter, because almost everywhere the limestone rocks rest upon mudstones that serve them as a base. I do not advance anything here that has not been demonstrated by experiment or confirmed by observations. All the world will be able to be assured, by procedures that are easy to repeat,[5] that glass and powdered sand transform over a short time into clay, solely by resting in water; and it is following this understanding that I have said, in my *Theory of the Earth*, that clays are only vitrescible sand that have decomposed and decayed. I can add here that it is probably due to this decomposition of vitrescible sand in water that one should attribute the origin of acid, because the principal acid that is found in clay can be regarded as a combination of vitrescible earth with fire, air, and water; and it is the same principal acid that is the main cause of ductility of clay and of all the other types of matter, without even excepting the bitumens, oils, and fats, which are only ductile and transmit ductility to other materials only because they contain acids.

After the fall and settling of boiling waters on the surface of the globe, the largest part of the glassy scoriae that entirely covered it was then converted in a fairly short time into clays; all the movements of the sea contributed to the prompt formation of these same clays, in stirring and transporting the scoriae and the glassy powders, and making them subject to the action of water from all sides. And a little later, the clays formed by the intermediary and effect of water were successively transported and deposited above the primitive rock of the globe, that is to say, above the solid mass of vitrescible matter that makes up the substrate and which, by its durability and hardness, resisted this same action of the waters.

The decomposition of these powders and vitrescible sands, and the production of clays, took place more quickly since the water was hot. This decomposition continued to take place and continues today, but more slowly and in lesser amounts, because as clays are present almost everywhere as a cover to the globe, and as these beds of clay are commonly a hundred or two hundred feet thick, and as the calcareous rocks and all the hills composed of these rocks generally rest on these beds of clay, one can sometimes find, below these same beds, vitrescible sands that have remained unconverted, and that conserve the character of the first origin. There are also vitrescible sands on the surface of the Earth and at the bottom of the sea, but the formation of these vitrescible sands present on the exterior date from a time long after the formation of these older sands of the same nature, which are found at great depths beneath the clays. Because these sands seen on the Earth's surface are only the detritus of granites, of sandstones, and of vitreous rock, masses of which form the cores and summits of the mountains, of which rain, ice, and other external agents have detached and continue to detach small parts every day, which were later entrained and deposited by running water at the Earth's surface. One thus must regard this production of glassy sands present at the bottom of the sea or the surface of the Earth as very recent by comparison with the other ones.

In this way the clays and the acid that they contain were produced a very short time after the establishment of the waters and a little time before the birth of shells—because we find in these same clays an infinite number of belemnites, of lenticular stones, of horns of Ammon, and of other specimens of these lost species of which we can nowhere find living analogues. I have myself found in a shaft which I had dug, at the lowest part of a small valley composed wholly of clay, and of which the neighboring valleys were also of clay up to eighty feet high—I have found, I affirm, belemnites that were eight inches long by almost an inch in diameter, and several of which were attached to a flat and thin part like the head of a crustacean. I have similarly found a great number of bronzed, pyritic horns of Ammon, and thousands of lenticular stones. These ancient remains were, as one can see, enclosed in clay at a depth of 130 feet; because although only dug fifty feet into this clay in the middle of the valley, it is certain that the thickness of this clay was originally 130 feet, since beds of it are elevated on both sides to eighty feet of height above. This was shown to me by the correspondence of these beds and by those of the banks of calcareous rock that lie above them on each side of the valley. These calcareous layers are fifty-four feet in thickness, and their different beds are found in corresponding positions and horizontally disposed at the same height above this immense bed of clay, which serves them as a base and which extends beneath the calcareous hills of all this country.

The time of formation of these clays thus immediately followed the establishment of the waters. The time of formation of these first shells must therefore be placed at a few centuries later, and the time of the transport of their remains followed almost immediately. There was no interval other than that which Nature placed between the birth and the death of these shelly animals. As the action of the waters each day converted the vitrescible sands into clays, and as its movement transported them from place to place, it entrained shells and other remains and debris of marine production, and deposited the whole as sediment, and formed the beds of clay in which we can find today these monuments, the most ancient of organized Nature, of which the models no longer survive. This is not to say that there are not also shells in these clays of which the origin is less ancient, and even some species that one can compare with those of our seas, and even better with those of southern seas. But this does not add any difficulty to our explanations, because water has not stopped converting into clay all of the glassy scoriae and vitrescible sands that are exposed to its action. It thus formed these clays in great quantity after it took possession of the Earth's surface. It continued and still continues to produce the same effect; because the sea today transports muds with the remains of presently living shells, just as it transported in other times the same muds with the remains of shells then living.

The formation of shales, slates, coals, and bituminous matter dates from about the same time: these materials are generally found in clays at some great depths. They seem even to have locally preceded the formation of the last beds of clay; because below the 130 feet of clay of which the beds contain belemnites, horns of Ammon, and other ancient shell debris, I have found carbonaceous and inflammable matter, and one knows that most coal seams are more or less overlain by beds of clayey earth. I think that I can even propose that it is in these earths that one should search for seams of coal of which the formation is a little more ancient than that of the external beds of clayey earth that overlie them. This is proved by these veins of coal being almost all inclined, while those of the clay, like all of the other external layers of the globe, are ordinarily horizontal. These last were thus formed by sediment of waters on a horizontal base, while the others, being inclined, seem to have been brought in by a current upon a sloping terrain. These beds of coal, which are all composed of plant remains more or less mixed with bitumen, owe their origin to the first vegetation formed by the Earth: all the parts of the globe that found themselves elevated above the waters produced from those first times an infinity of plants and trees of all species, which, soon toppling through old age, were entrained by the waters and formed deposits of vegetable matter in an infinity of places; and, like these bitumens and other terrestrial oils, seem to have been derived from animal and vegetable substances. At the same time, acid came from the decomposition of vitrescible sand by fire, air, and water, and finally acid entered into the composition of bitumens, because with vegetable oil and acid one can make bitumen. It seems that the waters were from then on mixed with these bitumens and were permanently impregnated, and as they incessantly transported trees and other vegetable matter that descended from the high ground of the Earth, these vegetable materials continued to mix with bitumens already formed from the residues of the first plants, and the sea, by its movement and by its currents, stirred, transported, and deposited on the clayey elevations that it had previously formed.

The beds of slate, which contain also plants and even fish, were formed in the same manner, and one can give examples that are so to speak before our eyes. Thus the slate and coal deposits were subsequently covered by other beds of clayey earth that the sea deposited in later times. There were even considerable intervals and changes in movement between the formation of different beds of coal in the same terrain; because one can often find, below the first bed of coal, a layer of clay or other deposit that follows the same inclination, and then one can find commonly enough a second bed of coal inclined like the first, and often a third, equally separated by layers of earth, and sometimes even by beds of calcareous rock, as in the coal seams of Haynault. One therefore cannot doubt that the lowest beds of coal were the first produced, by the transport of plant matter brought in by water. And, when the first deposit, from where the water carried away this plant matter, was exhausted, the movement of water continued to transport, to the same place, the soil or other matters that surrounded that deposit. These are the deposits that today form the intermediate layer between the two beds of coal, which suggests that the water then carried in some other deposit of plant matter to form the second bed of coal. I understand here by "beds" the entire layer of coal taken in all of its thickness, and not the small beds or laminae of the same substance of which coal is composed, and which are often extremely thin. These are the same laminae, always parallel to each other, that show that these masses of coal were formed and deposited by the sediment and even through the distillation of bitumen-impregnated waters; and these same types of laminae are found in new coals of which the beds are formed by distillation, at the expense of more ancient beds. Thus these laminae of coal took their form by two combined causes: the first is by deposition, always horizontal, from water; and the second, the disposition of plant matter, which tends to form laminae. There are, also, complete fragments of wood, commonly entire, and easily recognizable detritus of other plants, which clearly show that the substance of these earth-coals is only an assemblage of plant debris bound together by bitumen.

The single thing that can be difficult to conceive is the immense quantity of plant debris that these coal seams indicate, because they are very thick, very extensive, and found in an infinity of places. But if one pays attention to the perhaps even more immense production of vegetation that happens during twenty or twenty-five thousand years, and if one thinks at the same time that man had not yet been created: hence, there was no destruction of the vegetation by fire, and one will feel that it could not fail to be transported by water, and to form, in a thousand different places, very extensive beds of plant matter. One can imagine in a small way that which happened on a grand scale: what an enormous quantity of great trees, that some rivers, such as the

Mississippi, carry to the sea! The number of these trees is so prodigious that in certain seasons it hinders navigation on this great river. It is the same on the Amazon River and on most of the great rivers of deserted or sparsely inhabited continents. One can thus think, by this comparison, that all of the ground elevated above the waters was, at this beginning, covered in trees and other vegetation, which nothing destroyed except old age. There was made in this long period of time successive transportations of all these plants and their remains, entrained by water flowing from the high mountains to the seas. The same uninhabited countries of America can provide us another striking example: one sees in Guyana forests of the latanier palm, several leagues in extent, which grow in a type of swamp that one calls "drowned savannahs," which are nothing other than extensions of the sea. These trees, having lived through their span, fall of old age and are carried by the movement of water. The forests that are more distant from the sea and that cover all the high ground in the interior of the country are covered less by healthy and vigorous trees than by strewn trees that are decrepit and half decayed. Travelers who are obliged to pass the night in these forests need to take care to examine the spot they have chosen for camping, to see that they are surrounded only by solid trees, and they do not run the risk of being crushed as they sleep by the fall of some tree that is rotten to the core. The fall of these trees in great numbers is very frequent; a single gust of wind can often make a considerable toppling that one can hear the sound of from a great distance. These trees, rolling from the mountain heights, knock over a quantity of others, and they arrive together on lower ground, where they rot to form new beds of vegetal earth, or they are entrained by flowing waters into the neighboring seas, to form at this distance new beds of fossil coal.

The detritus of this plant matter is thus the main basis for coal seams: these are the treasures that Nature seems to have accumulated in advance for the needs to come of great populations. The more that humans multiply, the more the forests will diminish. The wood not being able to satisfy their consumption, they need recourse to these immense deposits of combustible matters, of which the use will become all the more necessary as the globe cools further; nevertheless they will never exhaust them, because a single coal seam perhaps contains more combustible matter than all the forests of a vast country.

Slate, which one can regard as hardened clay, is formed in beds, that similarly contain bitumen and plant remains, but in much smaller quantity; and at the same time they often contain shells, crustaceans, and fish that one cannot refer to any known species. Thus, the origin of the coal and of the slates dates from the same time; the only difference that there is between these two types of matter is that plants compose the greater part of the substance of coal, while the basic substance of slate is the same as that of clay, and plants, like fish, come to be found there only by chance and in small enough number. But, both contain bitumen, and were formed by laminae or by very thin beds always parallel to each other, which clearly shows that they also were produced as successive sediments in quiet water, of which the oscillations were perfectly regulated, such as are those of our ordinary tides or of constant water currents.

Thus taking again for a moment all that I have just presented, the mass of the terrestrial globe composed of molten glass, at first only showed blisters and irregular cavities that formed at the surface of all that matter that was liquefied by fire and of which parts were contracted by cooling. During this time and as cooling progressed, the elements were separated, and there was liquefaction and sublimation of metallic substances and minerals, and these occupied the cavities of high ground and the perpendicular fissures of the mountains. Because these highest points above the surface of the globe, having cooled first, also presented to the exterior elements the first fissures produced by the contraction of the matter that was cooling. The metals and the minerals were formed by sublimation, or deposited by waters in all these fissures, and it is for this reason that one finds nearly all of them in the high mountains, and one finds them in lower ground only as newly formed ores. A little while after the clays formed, the first shells and the first plants were born; and as they perished, their remains and their detritus formed calcareous rocks, and those of plants formed bitumens and coal; and at the same time the waters, by their movement and by their sediments, composed the organization of the surface of the Earth in horizontal beds; then the currents of these same waters gave it its exterior form by salient and re-entrant angles. And, it does not overly extend the time necessary for all these great operations and immense constructions of Nature, to count twenty thousand years since the birth of the first shells and the first plants. They were already much-multiplied and very numerous forty-five thousand years after the formation of the Earth: and as the waters, which at first were so prodigiously elevated, lowered successively and abandoned the regions that they had covered previously, these lands from then on presented a surface that was strewn with marine products.

The length of time during which the waters covered our continents was very long. This one cannot doubt in considering the immense quantities of marine products that are found to great enough depths and to very great heights in all the parts of the Earth. And how much time do we need to add, to this duration that is already so long, so that these same products can be broken, reduced to powder, and transported by the movement of water, and then to form marbles, calcareous rocks, and chalks! This long succession of centuries, this duration of twenty thousand years, seems to me still too short for the succession of effects that all these monuments show to us.

Because, one needs here to represent the march of Nature, and even to recall the idea of its mechanisms. Living organic molecules existed from the time when the elements, at gentle temperatures, could be incorporated with the substances that make up organized bodies. They produced, on the elevated parts of the globe, an infinity of plants, and in the waters an immense number of shells, of crustaceans, and of fish, which soon multiplied by means of generation. This multiplication of plants and of shells, as rapid as one can suppose it to have been, can have taken place only in a great number of centuries, because it produced volumes as prodigious as are those of their remains. In effect, to judge what has happened, one needs to consider what happens now. And so, does it not need many years for oysters, which live in some parts of the sea, to multiply in great enough quantity to form a type of rock? And how many centuries are needed to produce all the calcareous matter at the surface of the globe? And is one not forced to admit, not just centuries, but centuries of centuries so that the marine products were not just reduced to powder, but transported and deposited by the waters, in a manner that could form the chalks, the marls, the marbles, and the calcareous rocks? And how many centuries still need to be acknowledged for these same calcareous materials, newly deposited by the waters, to be purged of their superfluous humidity, then dried and hardened to the state that they are in today, and since so long ago?

As the terrestrial globe is not a perfect sphere, being thicker around the equator than beneath the poles, and as the action of the Sun is greater in tropical climates, it resulted that the polar countries were cooled sooner than those of the equator. These polar parts of the Earth thus first received the waters and volatile matters that fell from the atmosphere. The rest of these waters must then have fallen upon those climes that we call temperate, and those of the equator would have been last to be irrigated. There passed many centuries before the equatorial parts were sufficiently cooled to admit the waters. The equilibrium and even the occupation of the seas was long to form and be established; and these first inundations must have come from the two poles. But we have remarked that all of the terrestrial continents terminate in a point toward the southern regions. Thus, the waters come in greater quantity from the southern pole than the northern pole, from where they could only recirculate and not arrive, at least with such force. Without this, the continents would have taken a form completely different from that which they now show: they would be enlarged toward the southern shores rather than shrinking. In effect, the countries of the southern pole must have chilled more quickly than those of the northern pole, and by consequences received more water from the atmosphere, because the Sun stays a little less upon this southern hemisphere than upon the northern; and this cause seems to me sufficient to have determined the first movement of the waters and to have then perpetuated them for long enough to have shaped to a point the ends of all the terrestrial continents.

Moreover it is certain that the two continents were not separated toward our north, and even that their separation was only made long after the establishment of a living Nature in our northern climes, since the elephants existed at the same time in Siberia and Canada. This proves unarguably the continuity of Asia or Europe with America, whereas, by contrast, it seems equally certain that Africa was separated from America since the first times, since not one of the animals of the ancient continent has been found in this part of the New World, nor any remains that might indicate that they once existed there. It seems that the elephants of which bones have been found in North America were confined there, not being able to cross the high mountains to the south of the Isthmus of Panama, and they never penetrated into the vast regions of South America: but it is yet more certain that the seas that separate Africa and America existed before the birth of elephants in Africa. Because, if these two continents had been contiguous, the animals of Guinea would be found in Brazil, and one would have found the remains of these animals in South America as one finds them in the lands of North America.

Thus since the origin and at the beginning of a living Nature, the highest lands of the globe and the parts of our North were the first populated by the species of terrestrial animals best suited to great heat: the regions of the equator remained long as deserts, arid and without seas. The high lands of Siberia, of Tartary, and of other parts of Asia, all those of Europe that form the chain of mountains of Galicia, of the Pyrenees, of the Auvergne, of the Alps, of the Apennines, of Sicily, of Greece, and of Macedonia, like the mountains of the Himalayas and the Urals, and others, were the first inhabited countries, even over a number of centuries, while all the lower ground was still covered by the waters.

During this long interval of time when the sea rested on our lands, the sediments and the deposits of the waters formed the horizontal beds of the Earth: the lower ones of clay, and the upper ones of calcareous rock. It is in the sea itself that the petrifaction of marbles and of stone took place. At first, these materials were soft, having been successively deposited one above the other, in proportion to how the sea carried them in and how it let them fall in the form of sediments. Then, they were little by little hardened by the force of affinity of their constituent parts, and at last they formed all of the masses of calcareous rock, which are composed of horizontal beds or equally are inclined, as are all the other materials deposited by the waters.

It is since the first times of this same period that the clays were deposited where may be found the debris of ancient shells; and these shelled animals were not the only ones then existing in the sea. Because, apart from shells, one finds debris of crustaceans, of sea urchin spines, of sea lilies in these same clays. And in slates, which are only hardened clays mixed with a little bitumen, one finds, like in the shales, entire and well-preserved impressions of plants, of crustaceans, and of fish of different sizes. Finally, in coal seams, the entire mass of coal seems composed only of plant debris. These are the most ancient monuments of living Nature, and the first organized products both in the sea and on the land.

The northern regions, and the highest parts of the globe, and above all the summits of the mountains that we have listed, and which today mostly show only dry faces and sterile summits, were once fertile lands and the first ones where Nature was manifested. Because, these parts of the globe, having been sooner cooled than lower-lying lands or those closer to the equator, would have been the first to receive the waters of the atmosphere and all the other matter that could contribute to fertility. Thus one can presume that before the seas were fixed in place, all the parts of the Earth that found themselves above the waters were made fertile, and they could subsequently and over time produce the plants of which we find today impressions in the slates, and all of the vegetable substances that make up the Earth's coal.

In this same time when our lands were covered by the sea, and while the calcareous layers of our hills were formed of the detritus of these products, several monuments show us that a great quantity of vitrescible substances was detached from the summits of the primitive mountains and from the other uncovered parts of the globe, this being brought by alluviation, that is to say, by transport by waters, to fill the fissures and the other spaces left between the calcareous masses. These perpendicular or slightly inclined fissures in the calcareous layers were formed by the contraction of the calcareous matter, while they were dried and hardened, in the same manner in which were previously made the first perpendicular fissures in the vitrescible mountains produced by fire, as these materials contracted during their solidification. The rains, the winds, and other external agents had already detached a great quantity of small fragments from these vitrescible masses that the waters transported to different places. In searching for iron ore in the hills of calcareous rock, I have found many fissures and cavities filled with iron ore in grains, mixed with vitrescible sand and small rounded pebbles. These pods or nests of iron ore do not extend horizontally, but descend almost vertically, and they are all found at the crests of the highest of the calcareous hills. I have recognized more than a hundred of these pods, of which I have found eight principal and very considerable ones just in the extent of the terrain that neighbors my forges, one or two leagues away: all these ores were of generally small grains, and more or less mixed with vitrescible sand and small pebbles. I have effected the exploitation of five of these ores for use in my furnaces; some have been exploited to fifty or sixty feet, and others to a hundred and seventy-five feet of depth. All are sited in fissures in calcareous rocks, and there are in this region neither vitrescible rock, nor quartz, nor sand, nor pebbles, nor granite; and hence these iron ores, which form more or less large grains, and which are all more or less mixed with vitrescible sand and with small pebbles, could not have formed in the calcareous matter, where they are enclosed on all sides as within walls. By consequence they have all been brought from afar by the movement of waters that would have deposited them at the same time as they deposited the muds and other sediments; because these sacs of iron ore in grains are all overlain or laterally accompanied by a type of silky red earth, more workable, purer, and finer than common clay. It even seems that this silty earth, more or less colored with a red hue that iron gives to earth, is the ancient matrix of these iron ores, and it is in this same earth that the metallic grains must have formed before their transport. These ores, although situated in entirely calcareous hills, do not contain any gravel of the same type; there are, as one descends, only some isolated masses of calcareous stones around which are veins of the ore, always accompanied by red earth, which often traverses veins of the ore, or coats the walls of the calcareous rocks that enclose them. And, clearly proving that these deposits of ore are made by the movement of water, is that after emptying the fissures and cavities that contain them, one sees without a doubt, that the walls of these fissures have been affected and even polished by water, which consequently has filled and bathed them for a sufficiently long time before having deposited there the iron ore, the small pebbles, the vitrescible sand, and the silty earth, with which the fissures are now filled. And, one cannot be given to believe that the grains of iron were formed in this silty earth after they were deposited in these rocky fissures; because one thing as evident as the first opposes the idea, in that the quantity of iron ore seems to surpass greatly that of the silty earth. The grains of this metallic substance seem in truth to have all been formed in the same earth, which itself was produced only from the residue of animal and plant matter, in which we will demonstrate the production of iron in grains, though that was made before the transport and deposition in the rock fissures. The silty earth, the iron grains, the vitrescible sand, and the small pebbles were transported and deposited together; and if since then iron grains formed in this same earth, this was likely to have been only in small quantity. I have drawn thousands of casks from each of several mines and, without having exactly measured the quantity of silty earth that one has left in these same cavities, I have seen that it was much less than the quantity of iron ore in each.

But what proves again that these grains of iron ore were all brought in by the movement of water is that in the same canton, three leagues away, there is a fairly large extent of ground forming a small plain, above the limestone hills. These are as high as those I have just spoken about, and one finds in this terrain a great quantity of iron ore as grains, which is very differently mixed and situated. Because, instead of occupying perpendicular fissures and internal cavities in calcareous rocks, and instead of forming one or several perpendicular sacs, this iron ore is by contrast deposited *en nappe*, that is to say, as horizontal beds, like all water-laid sediment. Instead of descending deeply like the former, this extends almost to the surface of this terrain, with a thickness of several feet. Instead of being mixed with pebbles and vitrescible sand, it is by contrast mixed throughout only with gravel and calcareous sand. It shows one more remarkable phenomenon: this is a prodigious number of horns of Ammon and other ancient shells, so that it seems that the entire ore is composed of this. While, in the eight other ores of which I have spoken above, there is not the least vestige of shell, nor any fragment, any indication of any calcareous type, even though they are entirely enclosed within masses of rock that are entirely calcareous. This other ore, which contains such a prodigious amount of marine shell debris, even of the oldest ones, will thus have been transported with this shell debris, by the movement of water, and deposited in the form of sediment as horizontal beds; and the grains of iron that they contain, that are much smaller than those of the first ores, mixed with pebbles, would have been brought in with the shells themselves. Thus, the transport of all of this matter and the deposition of all of this iron ore in grains, was done by alluviation more or less at the same time, that is to say, while the seas still covered our limestone hills.

And the summits of all these hills, and of the hills themselves, no longer represent much of the aspect that they had when the waters abandoned them. As soon as their primitive form has been maintained, their salient and re-entrant angles have become more obtuse, their slopes less steep, their summits less elevated and more bare. The rains have loosened and entrained earth from them, and the hills have been lowered little by little, while the valleys have been filled with earth moved by rainwater or flowing water. One can imagine what would once have been the form of the terrain of Paris and its surroundings. In one part, on the hills from Vaugirard as far as Sève, one sees quarries in calcareous rocks full of petrified shells; on another side, near Montmartre, there are hills of plaster and of clay material. These hills, nearly equally elevated above the Seine, are today only of moderate height, but at the bottom of wells dug at Bissêtre* and the military school, human-worked wood was found at seventy-five feet of depth. Thus, one suspects that this valley of the Seine was filled with earth to a depth of seventy-five feet just since humans existed; and who knows how many of the adjacent hills were diminished over

* Now spelled Bîcetre, the ancient hospital in Paris.

the same time by the effect of rain, and what was the thickness of the earth with which they were once covered? It is the same with all the other hills and all the other valleys. They were perhaps twice as high and twice as deep at the time when the waters retreated and uncovered them. One can even be assured that the mountains are still being lowered day by day, and that the valleys are filling about in the same proportion. Only, this diminution of the height of the mountains, which takes place today in an almost imperceptible manner, took place much more quickly in these first times because of the greater steepness of their slopes, and it will now take several thousand years for the inequalities in the surface of the Earth to be reduced further, while this took place in a few centuries in the first ages.

But let us return to this epoch before the waters, after having arrived from the polar regions, reached those of the equator. It is in these terrains of the torrid zone that the greatest upheavals took place. To be convinced of this, one only has to cast one's eyes upon a geographical globe, and one will recognize that almost all of the space between the tropics of this zone shows only the debris of wrecked continents and of a ruined terrain. The immense quantity of isles, of straits, of heights and depths, of arms of the sea, and of intercut terrain, proves the numerous collapses that took place in that vast part of the world. The mountains there are higher and the seas deeper than anywhere else on Earth; and it is doubtless while these great collapses took place in the countries of the equator, that the waters that covered our continents were lowered and receded, flowing in great streams toward the lands of the south of which they filled the depths, first uncovering the highest parts of the lands and then all of the surface of our continents.

One can envisage the immense quantities of matter of all kinds that were then transported by the waters. How many sediments of different natures were then deposited one above another, and by consequence just how much of the first face of the Earth was changed by these revolutions? On the one hand, the flux and reflux gave to the waters a constant motion from east to west; on the other hand, the alluvium coming from the poles ran across this movement and directed the actions of the sea as much and perhaps more toward the equator than toward the west. How many individual inrushes took place from both sides? As each great subsidence event produced a new depth, the sea lowered and water rushed in to fill it; and although it seems today that the equilibrium of the seas is about settled, and that all of their action reduces to gaining some land toward the west and on letting it be exposed toward the east, it is nevertheless very certain that the seas every day lower their level more and more, and that they will be further lowered in proportion to some new subsidence, either by the effect of volcanoes or earthquakes, or by simpler and more constant causes: because all of the cavernous parts of the interior have not yet fallen in; volcanoes and earthquakes provide demonstrable proof. The waters little by little undermine the roofs and ramparts of these subterranean caverns, and while some of them collapse, the surface of the Earth being lost in these places, there will form new valleys that the sea will come to possess. Nevertheless as these events, which at the beginning must have been very frequent, are now quite rare, one can believe that the Earth has nearly arrived at a state quiet enough for the inhabitants to no longer fear the disastrous effects of these great convulsions.

The emplacement of all of these metallic materials and minerals has closely enough followed the establishment of the waters; that of clayey and calcareous matter has preceded their retreat; the formation, the situation, the position of all of these latter materials date from the time when the sea covered the continents. But one must observe that the general movement of the seas having started to take place as it does today, from east to west, they worked the surface of the Earth from east to west as much and perhaps more than they did previously, from south to north. One cannot doubt that if one pays attention to a very general and very true fact,[6] that in all of the continents of the world the slope of the ground, from the mountain summits, is always much more rapid on the western side than on the eastern side. This is evident in the entire continent of America, where the summits of the chain of the Cordillera neighbor very closely the seas of the west and are very distant to the sea of the east. The chain that separates Africa along its length, and that extends from the Cape of Good Hope to the Mountains of the Moon, is also closer to the seas of the west than of the east. There are similarly mountains that extend from the Comorin Cape in the Indian peninsula, and these are much closer to the sea in the west than that in the east. And if we consider the peninsulas, the promontories, the islands, and all the lands bordering the sea, we can recognize everywhere that the slopes are short and steep toward the west and are long and gentle toward the east. The backs of all of these mountains are similarly more scarped to the west than to the east, because the general movement of the seas is always from the east to the west, and as the waters were lowered, they destroyed the lands and eroded the backs of the mountains in the direction of their fall, as one sees in a cataract of eroded rocks and lands cut by the continual fall of water. Thus all of the terrestrial continents were first shaped to a point toward the south by the waters, which came from the southern pole more abundantly than from the northern pole; then they steepened the slope more rapidly in the west than the east in subsequent times, when these same waters obeyed the same general movement that carried them constantly from east to west.

FOURTH EPOCH

When the Waters Retreated and the Volcanoes Became Active

We have seen that the elements of the air and the water became established by cooling, and that the waters, at first consigned to the atmosphere by the expansive force of heat, then fell onto those parts of the globe that had cooled sufficiently so as not to drive them off as vapor, these parts being the polar regions and all the mountains. There was thus at the epoch of thirty-five thousand years ago a vast sea in the regions of each pole and some scattered lakes or ponds on the mountains and higher ground which, becoming cooled to the same degree as the polar regions, could equally receive and retain the waters. Then, as the globe cooled further, the seas around the poles, continually supplied by the fall of the waters from the atmosphere, extended further; and the lakes or large pools, similarly supplied by this incessant rain, which fell more abundantly as the cooling intensified, extended in all directions and formed basins and small interior seas in those parts of the globe that the great seas of the two poles had not yet reached. Then the waters, continuing to fall in ever greater volume until the complete purification of the atmosphere, spread successively over the terrain to reach the regions of the equator. Eventually, they covered all of the surface of the globe to two thousand toises higher than the level of our present-day seas. The entire Earth was then subject to the empire of the sea, with the possible exception of the summits of the primitive mountains, which were only, so to speak, washed and bathed during that first time when the waters fell, to flow from these high areas to occupy lower ground as soon as this had cooled sufficiently to hold them without driving them off as vapor.

Thus a universal sea formed, which was only interrupted and overlooked by the mountain summits from where the first waters had already departed in flowing to lower ground. These high areas, having first been acted upon by the presence and movement of the waters, would also have been the first to be made fertile; and as all of the surface of the globe was only, so to speak, a kind of archipelago, organized Nature established itself on these mountains, and it even spread there with great vigor; because the heat and humidity, these two principles of all fertility, were there united and combined to a greater degree than they are today in any climate of the Earth.

And in these same times when the lands raised above the waters became covered with great trees and vegetation of all kinds, the worldwide sea was filled everywhere with fish and shellfish; it was also the universal repository for all that became detached from the lands that overlooked it. Scoriae of primitive glass and vegetation were carried from the heights of the land to the depths of the sea, at the bottom of which they formed the first beds of vitrescible sand, of mud, of shale, and of slate, together with the first seams of coal, of salt, and of bitumens, which from then impregnated all of the mass of the seas. The quantity of vegetation produced and destroyed in these first lands is too immense to be represented; because when we reduce the surface of all these lands that were then raised above the waters to a hundredth or even a two-hundredth part of the surface of the globe, that is to say to a hundred and thirty thousand square leagues, it is easy to sense how many trees and plants this vast terrain of one hundred and thirty thousand square leagues produced over several thousand years, how much of such detritus was accumulated, and in what enormous quantity it was entrained and deposited beneath the waters, where it formed the foundation for the quantity of coal seams that are found in so many places. There are similarly layers of salt, seams of iron ore, of pyrites, and of all the other substances with a composition in which acids have played a role and of which the first formation could take place only after the fall of the waters. These materials would have been entrained and deposited in the low places and in the fissures of the rocks of the globe, where, finding already-formed mineral substances sublimated by the great heat of the Earth, they would have formed the first basis for the fuel of the volcanoes that were to come. I say "to come" because no active volcanoes existed before the waters had become established, and they did not begin to be active or rather could take on permanent activity only after the water level lowered. Because, one has to distinguish terrestrial volcanoes from marine volcanoes; the latter can only produce explosions that are, so to speak, momentary, because at the instant that their fire is ignited by effervescence of pyrite and combustible materials, it is immediately extinguished by the water that covers them and streams and gushes toward their furnaces, by all the routes that that fire opens to escape. The terrestrial volcanoes have by contrast an action that is durable and proportional to the quantity of materials that they contain; these materials have need of a certain quantity of water to enter a state of effervescence, and it is therefore only by the shock of a large amount of fire meeting a great volume of water that they can produce their violent eruptions. Similarly, while a submarine volcano can be active only for instants, a terrestrial volcano can maintain its activity only while it is in the vicinity of water. It is for this reason that all the volcanoes active today are on islands or near the sea's coast, and that one can list a hundred times more extinct volcanoes than active ones; because as the waters, in retreat, become too distant from the foot of these volcanoes, their eruptions diminish by degrees and finally cease entirely, and the gentle effervescence that rainwater can cause in their ancient hearth can produce a detectable effect only in particular and very rare circumstances.

These observations confirm perfectly what I say here on the action of volcanoes: all those that are active now are situated close to the sea. Those that are extinct, and of which the number is much greater, are sited in the centers of lands, or at least at some distance from the sea; and while most volcanoes seem to belong among the highest mountains, there exist many others among peaks of moderate height. The age of these volcanoes is thus not the same everywhere: first, it is certain that the first of them, that is to say the most ancient ones, could become permanently active only after the lowering of the waters that covered their summit; and subsequently, it seems that they ceased to be active once these same waters became too distant from their vicinity: because, I repeat, no power, with the exception of a great mass of water being shocked by a large amount of fire, could produce movements as prodigious as those of the eruption of volcanoes.

It is true that we do not see closely enough the interior composition of these terrible mouths of fire, to be able to pronounce on their effects with perfect knowledge of their cause. We only know that there are often subterranean communications from volcano to volcano. We also know that, although the source of their heat is perhaps not a great distance from their summit, there are nevertheless cavities that descend much lower, and that these cavities, the depth and extent of which are unknown to us, could be all or partly filled with the same materials that are presently heated.

Taking another view, it seems to me that electricity plays a very great role in earthquakes and in volcanic eruptions. I am convinced by very solid reasons, and by the comparison that I have made of experiments on electricity, *that the basis of electrical matter is the inherent heat of the terrestrial globe.* The continual emanations of this heat, although detectable, are not visible, and stay in the form of obscure heat while they have free and direct movement. But they produce a very bright fire and powerful explosions, once they are deflected from their direction, or accumulated by the friction of bodies. The interior cavities of the Earth containing fire, air, and water, the action of this first element must produce violent winds, loud storms, and subterranean thunder, of which the effects can be compared to lightning in the air. These effects must even be more violent and longer-lasting, because of the strong resistance that the solidity of the Earth opposed on all sides to the electrical force of this underground thunder. The buoyancy of air mixed with dense vapors and inflamed by electricity, the work of water, turned to elastic vapor by fire, and all the other impulsions of this electrical force raise and open the surface of the Earth, or at least agitate it by shaking, of which the tremors do not last much longer than the internal lightning that produced them. And, the tremors renew themselves until the expanding vapors issue from some opening at the surface of the Earth or within the heart of the seas. The eruptions of volcanoes and earthquakes are also preceded and accompanied by a dull and rolling noise, which only differs from that of thunder by its sepulchral and profound tone that this sound necessarily takes in traversing a great thickness of solid matter in which it remains enclosed.

This subterranean electricity, combined as a general cause with particular causes of fires lit by the effervescence of pyritic and combustible matter that the Earth conceals in so many places, suffices to explain the principal phenomena of the action of volcanoes. For example, their center seems to be fairly close to their summit, but the storm forms beneath. A volcano is only a vast oven, in which the bellows, or rather the ventilators, are sited in the interior cavities, to the side and below their center. These are the same cavities, since they extend to the sea, that serve as suction pipes to carry upward, not just vapors, but even masses of water and air. It is during this transport that is produced the subterranean lightning, announced by roaring, and only explodes by the dreadful vomiting of materials that it had broken, burnt, and calcined. There are thick billows of black smoke or of sullen flame; heavy clouds of ashes and stones; boiling torrents of molten lava, spreading far their burning and destructive flows—and extending far beyond the convulsive movements in the entrails of the Earth.

These internal tempests become all the more violent as they near volcanic mountains and the waters of the sea, of which the salt and fatty oils further increase the intensity of the fire; the ground situated between the volcano and the sea cannot fail to experience frequent shocks. But why is there no place in the world where we have not felt, even in living memory, some trembling, some vibration caused by the internal movements of the Earth? They are in truth less violent and much rarer in the middle of continents, distant from volcanoes and from the sea; but do not these effects depend upon the same causes? Why then are they felt there where there are no such causes, that is to say, in those places where there are neither seas nor volcanoes? The answer is easy—and that is there were seas everywhere and volcanoes nearly everywhere, and although their eruptions stopped when the seas became more distant, their fires persist, and are shown to us by the sources of terrestrial oils, by the hot and sulfurous fountains that are frequently found at the foot of the mountains, as far as the centers of the largest continents. These fires of the ancient volcanoes, become more tranquil since the retreat of the waters, suffice nevertheless, from time to time, to excite internal movements and to produce a gentle shaking, of which the oscillations are directed in the sense of the cavities of the Earth, and perhaps in the waters or of the veins of metal, as conductors of this subterranean electricity.

One could also ask me, why are all the volcanoes situated in the mountains? Why do they seem to be fiercer as the mountains become higher? What is the cause that could arrange these chimneys in the interior of the most solid walls, the most elevated on the globe? If one has understood well what I have said on the subject of irregularities produced by the first cooling, when the molten matter became solid, one will sense that the high mountains represent the largest blisters that were made on the surface of the globe at the time when it acquired its solidity. Most of the mountains are therefore situated above cavities, against which terminate the perpendicular fissures that cut them from high to low: these caverns and fissures contain materials that become inflamed solely by effervescence, or that are set alight by electrical sparks from the internal heat of the globe. Since the fire began and made itself felt, the air drawn in by rarefaction increased its force and soon produced a great blaze, the effect of which is to produce in turn movements and internal storms, the subterranean thunder and all the impulsions, the noises and shocks that precede and accompany the eruptions of volcanoes. One thus must stop being astonished that volcanoes are all situated in high mountains, because these are the only ancient places of the Earth where the interior cavities are maintained, the only ones where the cavities interconnect from bottom to top by fissures that are not yet infilled, and finally they are the only ones where the empty space is vast enough to contain the very great quantities of materials that serve as the source of the fire of permanent and still-active volcanoes. Besides, they will become extinct like the others as the centuries pass; the eruptions will cease: dare I even say that humans could contribute to that? Would it cost as much to cut the communication between a volcano and its neighboring sea, as it cost to construct the pyramids of Egypt? These useless monuments of a false and vain glory at best teach us that in employing the same forces for the monuments of wisdom, we can do very great things, and perhaps even control Nature, to the point of stopping, or at least of directing, the ravages of fire, just as we already know how to direct and break the effects of water.

Up until the time when volcanoes became active, there existed only three types of matter on the globe: (1) the vitrescible matter produced by primitive fire; (2) the calcareous matter produced by the intermediary of water; (3) all the substances produced from the detritus of animals and of plants. But, the fire of the volcanoes gave birth to matter of a fourth kind that often participates in the nature of the three others. The first class includes not only the first solid and vitrescible materials, of which the nature has not been altered at all, which form the foundations of the globe and also the cores of the primordial mountains, but also the sands, shales, slates, mudstones, and all the vitrescible materials that have been decomposed and transported by the waters. The second class contains all the calcareous materials, that is to say all of the substances produced by shellfish and other animals of the sea. They extend across entire provinces and cover reasonably large countries; they are found also at considerable depths, and they surround the bases of the highest mountains up to a very large height. The third class comprises all of the substances that owe their origin to animal and plant matter, and these substances occur in very great amounts; their quantity seems immense, because they cover all of the surface of the Earth. Finally, the fourth class is that of materials brought up and expelled from volcanoes, of which some appear to be a mixture of the first class, and others unmixed have undergone a secondary action by fire that has given them a new character. We place within these four classes all of the mineral substances, because on examining them, one can always distinguish which of these classes they belong to, and consequently can pronounce on their origin. This suffices to show us the approximate time of their formation; because, as we have demonstrated, it seems clear that all of the solid vitrescible materials that have not changed their nature or situation have been produced by primitive fire, and so their formation belongs to our second epoch; while the formation of the calcareous materials, and that of mudstone, coal, and so on took place only in subsequent times and can be placed in our third epoch. And as, in the matter expelled from volcanoes, one can sometimes find calcareous substances and often sulfur and bitumens, one can hardly doubt that the formations of these substances thrown out of volcanoes postdates the formation of all of these materials and belongs to our fourth epoch.

Although the quantity of matter thrown out by volcanoes is very small by comparison with the quantity of calcareous materials, it nevertheless covers large enough extents on the surface of lands situated near these fiery mountains and those of which the fires are extinct and stilled. By their repeated eruptions, they filled valleys, covered plains, and even produced other mountains. Then, when the eruptions ceased, most of the volcanoes continued to burn, but with a gentle flame that did not produce any violent explosion because, being distant from the sea, there was no longer the shock of water against fire; the effervescent materials and the combustible substance set alight long ago continued to burn, and it is what today produces the heat of all our thermal waters; they pass over the hearths of these underground fires and come, very hot, out of the heart of the Earth. There are also several examples of coal seams that burn from times immemorial, and that were lit by underground lightning or by the gentle fire of a volcano that has ceased erupting. These thermal waters and burning coal seams are often found, like the extinct volcanoes, in regions far from the sea.

The surface of the Earth shows us, in a thousand places, the vestiges and the proofs of these extinct volcanoes. In France alone, we know the old volcanoes of the Auvergne, of Velai, of Viverais, of Provence, and of Languedoc. In Italy, almost all of the ground is formed of the debris of volcanic matter, and it is the same in several other countries. But to unite these objects in a general point of view, and to clearly conceive the order of the upheavals that the volcanoes produced on the surface of the globe, one needs to go back to our third epoch at the time when the sea was universal and covered all of the surface of the globe with the exception of the high places where the first mixture of vitreous scoriae of the terrestrial mass with the waters was made. It is at this very same date when the vegetation was born and multiplied on those lands that the sea had just abandoned. The volcanoes did not yet exist, because those materials that served as fuel for their fire, that is to say, the bitumens, the coals of the earth, the pyrites, and even the acids, could not have previously formed, as their composition presupposed the intermediary of water and the destruction of vegetation.

Thus the first volcanoes existed on the high ground in the middle of the continents, and as the sea withdrew and became more distant from their foot, their fires died down and ceased to produce the violent eruptions, which could operate only by the conflict of a great mass of water against a great volume of fire. And it took twenty thousand years for this successive lowering of the seas and for the formation of all our limestone hills. And, as the pile of combustible matter and minerals that served to fuel the volcanoes could be deposited only successively, much time must have elapsed before they could be made active. It was only barely at the end of that period, that is to say, fifty thousand years from the formation of the globe, that the volcanoes began to ravage the Earth. As the surroundings of all the exposed land were still bathed with water, there were volcanoes almost everywhere, and there were frequent and prodigious eruptions that ceased only after the retreat of the waters. But, since this retreat could take place only after the subsidence of the blisters of the globe, it often happened that the water, streaming in floods to fill the depths of the subsided land, activated the submarine volcanoes that, by their explosions, pushed up part of those newly subsided lands, and sometimes pushed them above the level of the sea, where they formed new islands, as we have seen in the small island that formed beside that of Santorini. Nevertheless, these effects are rare, and the action of the submarine volcanoes is neither permanent nor powerful enough to raise up a large extent of land above the surface of the seas. The terrestrial volcanoes, by the continuation of their eruptions, in contrast covered all the areas that surrounded them with their rubble. By the successive deposition of their lavas, they formed new beds. These lavas that became fertile with time are an irrefutable proof that the primitive surface of the Earth, at first molten, then solidified, could likewise become fertile. Finally, these volcanoes also produced knolls, or mounds, which are seen on all volcanic mountains, and they built up ramparts of basalt, which can serve as the shorelines of neighboring seas. Thus, after the water, by its uniform and constant movement, produced the horizontal construction of the layers of the Earth, the fire of the volcanoes, by its sudden explosions, disrupted, cut, and buried some of these beds; and one cannot be surprised to see matter of all kinds emerging from the heart of the volcanoes, cinders, calcined stones, burnt earth, nor to find these materials mixed with calcareous and vitrescible substances, of which these same beds are composed.

The earthquakes could be felt long before the eruptions of the volcanoes; from the first moments of the subsidence of the caverns, there were violent shocks, which produced effects that were just as violent and much more prolonged than those of the volcanoes. To give some idea of this, let us suppose a cavern holding up terrain of one hundred square leagues, which would comprise only one of the small blisters on the surface of the globe, would suddenly collapse. Would not this collapse have necessarily been followed by a commotion that would be communicated and experienced for a great distance by more or less violent shaking? Although a hundred square leagues make up only one two hundred sixty thousandth of the Earth's surface, the fall of this could not fail to jolt all of the adjacent lands, and perhaps to simultaneously cause the collapse of neighboring caverns. There was thus no considerable subsidence that was not accompanied by the violent shocks of earthquakes, of which the movement was communicated by the elastic force that all matter possesses, and that must sometimes be propagated very far by the routes that can be offered by the voids of the Earth, in which the subterranean winds, excited by these commotions, could perhaps light the fires of the volcanoes, with the result that a single cause, that is to say the collapse of a cavern, can give rise to several effects, all great, and for the most part terrible. (1) At first, the lowering of the sea, forced to race in great floods to fill this new deep and consequently to leave new terrains exposed. (2) The shaking of the neighboring ground, through the commotion of the fall of the solid materials that formed the vaults of the cavern; and this shaking made the mountains tilt, split them to their summit, and detached from them masses that rolled to their base. (3) The same movement produced by the commotion, and propagated by the winds and underground fires, raised far-off ground and water, elevated the knolls and mounds, formed chasms and crevasses, changed the course of rivers, stopped up ancient sources and produced new ones, and ravaged, in less time than I take to recount it, all that found itself in its vicinity. We have thus to cease being surprised to see, in so many places, the horizontal uniformity of the work of the waters broken and cut by inclined fissures, with irregular landslides, and frequently hidden by inchoate rubble accumulated randomly, any more than to find such great terrains altogether covered with materials expelled from volcanoes. This disorder caused by earthquakes, nevertheless, only masks Nature to those who see it only on a small scale and who, from an accidental and particular effect, make a general and constant cause. It is water solely that, as a general cause subsequent to that of primitive fire, managed to construct and shape the present-day surface of the Earth; and that which lacks in uniformity in this universal construction is only the particular effect of an accidental cause by earthquakes and of the action of volcanoes.

Now in this construction of the surface of the Earth, by the movement and the sediment of the waters, one needs to distinguish two periods of time. The first began after the establishment of the universal sea, that is to say, after the perfect clearing out of the atmosphere, by the fall of the waters and of all the volatile matter that the heat of the globe had held aloft. This period lasted as long as it was necessary to multiply the shellfish, to the point of filling all of our limestone hills with their remains, and as long as it was necessary to multiply the vegetation, and to form from their debris all of our coal seams; and finally as long as was needed to convert the scoriae of primitive glass into muds, and to form the acids, the salts, the pyrites, and so on. All of these first and great effects were produced together during the time that passed since the establishment of the waters up to their lowering. Then the second period commenced. This retreat of the waters did not happen suddenly, but over a long succession of time, in which one needs to distinguish some separate events. The mountains composed of calcareous rock were certainly constructed in this ancient sea, of which the different currents certainly shaped them along corresponding angles. And close inspection of the sides of our valleys shows us that the *particular work of the currents was subsequent to the general work of the sea*. This fact, which one did not even suspect, is too important to not support all that could be rendered visible to all eyes.

Let us take for example the highest limestone mountain in France, that of Langres, which rises above all of the terrains of Champagne, and extends in

Bourgogne as far as Montbard, and even to Tonnerre, and which in the opposite direction similarly dominates the terrains of Lorraine and the Franche-Comté. This continuous cordon of mountains of Langres which, from the sources on the Seine as far as those of the Saône, is more than forty leagues in length, is entirely calcareous, that is to say, entirely composed of marine products; and it is for this reason that I have chosen it to serve as example. The highest point of this chain of mountains lies very close to the town of Langres, and one can see that, on one side, this same chain sends its waters into the ocean via the Meuse, the Marne, the Seine, and so on, and from the other side, it supplies them to the Mediterranean by the rivers that end in the Saône. The point where Langres is situated is found about in the middle of this length of forty leagues, and the hills descend roughly equally toward the sources of the Seine and toward those of the Saône. Finally, these hills, which form the extremities of this chain of limestone mountains, both end in terrain of vitrescible matter: beyond the Armanson near the Sémur, on the one hand; and beyond the sources of the Saône and of the little river of Conay, on the other.

In considering the little valleys next to these mountains, we can establish that the point of Langres, being the highest, was the first exposed when the waters receded. Before, this summit had been covered like all the others by the waters, because it is composed of calcareous materials. But at the moment when it became exposed, the sea was no longer able to overtop it, and all of its movements were reduced to pound this summit from two sides, and consequently to cut, by its constant currents, the gullies and the valleys that today follow the streams and rivers that flow down the two sides of these mountains. The evident proof that the valleys were all cut by constant and regular currents is that their salient angles correspond everywhere to their re-entrant angles; one only observes that the waters, having followed the steepest slopes, only broaching at first the less solid ground and that which is easiest to partition, there is often a remarkable difference between the two sides of the valley. One sometimes sees a considerable escarpment and jagged rocks on one side, while on the other, the banks of stone are covered by ground of a gentle slope; and this necessarily happens every time when the force of the current is focused more on one side than on the other, and also every time it was impeded or assisted by another current.

If one follows the course of a river or stream adjoining the mountains from their sources, one will easily recognize the shape and even the nature of the ground that forms the sides of the valley. In those places where it is narrow, the direction of the river and the angle of its course indicates at first glance the side toward which it carries its waters, and consequently the side where the ground forms a plain, while on the other side it continues to be in the mountains. Where the valley is large, this judgment is more difficult; nevertheless one can, in following the direction of the river, work out well enough on which side the ground widens or narrows. That which the rivers do today in a small way, the currents of the sea once did in a greater manner. They carved all our valleys and cut them from both sides, but in transporting the rubble, they often formed escarpments on one side and plains on the other. One must also remark that in the vicinity of the summit of these limestone mountains, and particularly at the summit of Langres, the little valleys begin with a circular hollow, and that from there they always flow and become larger as they become more distant from their source. These little valleys also seem deeper at the point where they begin and seem always to diminish in depth as they become wider and more distant from this point. But this is an appearance rather than a reality, because during its origin, the part of the valley nearest to the summit was the narrowest and the least deep. The movement of the waters began by forming a ravine, which became enlarged and excavated little by little. The rubble, having been entrained and transported by the currents of water to the lower parts of the valley, would have filled its bottom, and it is for this reason that the valleys seem deeper at their origin than over the rest of their course, and that the large valleys seem to be less deep as they become more distant from the summit against which their branches terminate. Because, one can consider a large valley as a trunk that throws off branches as other valleys, which throw off twigs as other little valleys, which extend and climb as far as the summit against which they terminate.

In following this object in the example we have just given, if one takes together all the terrains that pour their waters into the Seine, this vast space will form a valley of the first order, that is to say, of the greatest extent. Then, if we take only those lands that carry their waters to the river Yonne, this space will be a valley of the second order, and continuing to climb toward the summit of the chain of mountains, these lands that pour their waters into the Armançon, the Serin, and the Cure form valleys of the third order, and then the Brenne, which falls into the Armançon, will be a valley of the fourth order, and finally the Oze and the Ozerain, which fall into the Brenne, and of which the sources are next to those of the Seine, form valleys of the fifth order. It is the same if we take the terrains that carry their waters to the Marne; this space will be a valley of the second order. And, continuing to climb to the summit of the chain of mountains of Langres, if we take only the terrains of which the waters flow into the Rognon river, that will be a valley of the third order. Finally, the terrains that pour their waters into the streams of Bussière and Orguevaux form valleys of the fourth order.

This disposition is general in all of the terrestrial continents. As one climbs and approaches the summit of chains of mountains, one clearly sees that the valleys are narrower; but, although they also seem deeper, it is nevertheless certain that the ancient floor of the lower valleys was much deeper formerly than are those of the upper valleys today. We have said that in the valley of the Seine in Paris, wood worked by the hand of man has been found at seventy-five feet of depth; the first floor of this valley was thus in another time much lower than it is today, because below these seventy-five feet one must yet find rocky and terrestrial debris entrained by currents from the general summit of the mountains, as much from the valleys of the Seine as from those of the Marne, the Yonne, and all the rivers that they receive. By contrast, as one digs in the little valleys of the general summit, one does not find any rubble, but solid banks of calcareous rock seen as horizontal beds, with mudstones below at a more or less great depth. I have seen in a gorge near enough to the crest of this long cordon of the mountain of Langres, a well of two hundred feet of depth dug into the calcareous rock before finding mudstone.

The initial floor of the great valleys formed by primitive fire or even by the currents of the sea has thus been successively covered and built up by all the volume of the rubble entrained by the currents as it eroded the higher ground. The floor of the higher valleys has remained almost bare, while those of the lower valleys has been filled with all of the matter that the others have lost; such that when one sees only superficially the surface of our continents, one can fall into the error of dividing them into bands that are sandy, marly, shaley, and so on. Because, all these bands are only some superficial debris that prove nothing, and which, as I have said, do nothing to mask the true nature and deceive us as to the true theory of the Earth. In the higher valleys, one cannot find other debris than that which has descended by the effect of pluvial water long after the retreat of the seas, and this debris has formed the little beds of earth that presently cover the floor and the sides of these small valleys. This same effect has taken place in the large valleys, but with this difference that in the small valleys, the soils, the gravels, and the other detritus brought in by the pluvial waters and by the streams were deposited immediately upon a bare floor and then swept by the currents of the sea, while in the large valleys, this same detritus brought in by the pluvial waters could only be superposed on much thicker beds of debris previously entrained and deposited by these same currents. It is for this reason that, over all the plains and in the great valleys, our prior Observers of Nature thought to have found Nature in disorder, because they saw there calcareous materials mixed with vitrescible materials, and so on. But is this not to judge a building by its rubbish, or any other construction by chips of its materials?

Thus, without dwelling on these narrow and false views, let us follow our object in the example that we have given.

The three great currents that have formed beneath the summit of the mountain of Langres are represented to us today by the valleys of the Meuse, of the Marne, and of the Vingeanne. If we examine these terrains in detail, we will see that the sources of the Meuse came in part from the swamps of Bassigny, and from other small very narrow and scarp-like valleys; that the Mance and the Vingeanne, which both flow into the Saône, also come out of the very narrow valleys on the other side of the summit; that the valley of the Marne below Langres is about three hundred toises in depth; that in all of the first small valleys, the sides neighbor each other and are scarp-like; that in the lower valleys, and as the currents become more distant from the general and common summit, they become great in extent, and have in consequence enlarged the valleys, of which the sides are also less scarp-like, because the movement of the waters there was freer and less rapid than in the narrow small valleys of the terrains neighboring the summit.

One must also remark that the direction of the currents has changed in their course, and that the declination of the sides has changed for the same reason. The currents that descended toward the south, and that are represented by the little valleys of the Tille, of the Venelle, of the Vingeanne, of the Saulon, and of the Mance, acted more strongly against the sides facing the summit of Langres and with a northern aspect. The currents by contrast that descended to the north, and that are represented to us by the small valleys of the Aujon, of the Suize, of the Marne, and of the Rognon, like those of the Meuse, acted more strongly against the sides that face the same summit of the Langres, which are found with a southern aspect.

There was thus, while the waters exposed the summit of Langres, a sea of which the movements and the currents were directed toward the north, and on the other side of this summit, another sea whose movements were directed towards the south. These two seas beat against the opposing flanks of this chain of mountains, as one sees in today's sea the waters beat against the opposing flanks of a long island or of a promontory. It is therefore not astonishing that all the scarp-like sides of these valleys are equally found on the two sides of this general mountain summit; this is only the necessary effect of a very evident cause.

If one considers the terrain that surrounds one of the sources of the Marne near Langres, one will recognize that it emerges from a semicircle cut almost vertically; and in examining the beds of stone of this type of amphitheater, one will demonstrate that those of the two sides and those of the floor of the arc of the circle that it presents were once continuous and formed a single mass, which the waters have destroyed in the part that today forms this semicircle. One will see the same thing at the origin of the two other sources of the Marne, knowing that, in this small valley of Balesme and in that of Saint-Maurice, all of this terrain was continuous before the retreat of the sea; and this kind of promontory, at the extremity of which the town of Langres is sited, was at the same time continuous, not only with these first terrains, but also with those of Breuvonne, of Peigney, of Noidan-de-Rocheux, and so on. It is easy to convince oneself, using your own eyes, that the continuity of these terrains was destroyed only by the movement and action of water.

In this mountain chain of Langres, one finds several isolated hills, some in the form of truncated cones, like that of Montfaugeon, others with an elliptical form, like those of Montbard, of Montréal; and others just as remarkable, around the sources of the Meuse, toward Clemont and Montigny-le-Roi, which is situated on a mound adhering to the continent by a very narrow tongue of land. One sees another of these isolated hills at Andilly, and another by d'Heuilly-Coton, and so on. We must observe that in general these isolated limestone hills are lower than those that surround them, and from which these hills are presently separated, because the current, filling all the width of the valley, directly passed over these isolated hills and destroyed them at the summit, while it only bathed the ground of the valley sides, and only attacked them through an oblique motion, resulting in the mountains that border the valleys remaining higher than the isolated hills that lie between them. At Montbard, for example, the height of the isolated hill on top of which are sited the walls of the old chateau is only one hundred and forty feet, while the mountains that border the little valley on two sides, to the north and to the south, are more than three hundred and fifty feet high. It is the same with the other limestone hills we have just cited; all those that are isolated are at the same time less high than the others, because being in the middle of the valley and in the line of the current, they were eroded at their summits by the current, always more violently and quickly in the middle than at the sides of its course.

When one regards the escarpments, often several hundred toises high at their pinnacles, while one sees them made up from top to bottom of very massive and very hard banks of calcareous rock, one is amazed by the prodigious time that one would think needed for the water to have opened and excavated these enormous trenches. But, two circumstances competed in the acceleration of this great work. One of the circumstances is that in all the hills and calcareous mountains, the higher beds are the least compact and softest, so that the waters easily cut into the surface of the ground and formed the first ravine that then directed their course. The second circumstance is that, although the banks of calcareous rock were formed, and even dried and petrified, below the waters of the sea, it is nevertheless very certain that they were initially only superposed sediments of soft materials, and they acquired their hardness only progressively by the action of gravity on the total mass, and by the exercise of the force of affinity of their constituent parts. We are therefore assured that these materials had not acquired all the solidity and hardness that we see today, and that in the time of activity of the sea currents, they could yield to them with less resistance. This consideration diminishes the enormity of the duration of the work of the waters, and better explains the correspondence between the salient and reentrant angles of the hills, which perfectly resemble the correspondence of the river banks in all of the terrains that are easy to divide.

It is for the very construction of these limestone terrains, and not for their separation, that is it necessary to admit a very long period of time; and so in the twenty thousand years, I have taken at least the first three-quarters for the multiplying of the shellfish, the transport of their remains, and the formation of the masses that enclose them, and the last quarter for the separation and for the configuration of these same calcareous terrains. Twenty thousand years were needed for the retreat of the waters, which at first were raised by two thousand toises above the level of our present-day seas. And, it is only toward the end of this long march of retreat that our valleys were cut, our plains established, and our hills exposed. During all of this time the globe was inhabited by only fish and shelled animals. The mountain summits, and some high ground that the waters did not submerge, or that they abandoned early, were also covered by vegetation; because their detritus, in immense volumes, formed the seams of coal, at the same time as the remains of shellfish formed the beds of our limestone rocks. This is thus demonstrated by close attention to these authentic monuments of Nature; to know that the shells in the marbles, the fish in the slates, and the plants in the coal seems were all organized beings that existed long before the terrestrial animals, especially as we find no trace, no vestige of their existence of those animals in all the ancient beds that were formed from the sediment of the waters of the sea. One has found the bones, the teeth, the tusks of terrestrial animals only in the superficial layers, or in the valleys, or on the plains of which we have spoken, which were built up of detritus entrained from the higher ground by flowing water. There are only a few examples of bones found in the cavities below the rocks, near the edge of the sea, and in the low-lying terrains. But, these rocks beneath which are scattered these terrestrial animal bones are themselves newly formed, as are all the limestone deposits in the low countries, which are formed only of the detritus of ancient beds of rock, all situated above these new deposits. It is for this reason that

I have distinguished them by the name *parasite deposits*, because they are indeed formed at the expense of the first ones.

Our globe, over thirty-five thousand years, was thus only a mass of heat and fire, which no sensitive being could approach. Then, during fifteen or twenty thousand years its surface was only a universal sea. This long succession of centuries was needed for the cooling of the Earth and for the retreat of the waters, and it was only at the end of this second period that the surface of our continents was shaped.

But these last effects of the action of water were preceded by some other effects that were still more general, and these influenced in some ways the entire face of the Earth. We have said that the waters, coming in greater quantity from the southern pole, had sharpened all of the termini of the continents. But, after the complete fall of the waters when the universal sea had assumed its equilibrium, the movement from south to north ceased, and the sea had no more than the constant power of the Moon to obey which, combining with that of the Sun, produced the tides and the constant movement from east to west. The waters, as they first arrived, were first directed from the poles to the equator, because the polar parts, which had cooled more than the rest of the globe, received them the first. Then, they successively reached the regions of the equator, and while these regions were covered like all the others by the waters, the movement from east to west was from then established forever; because not only was it maintained during this long period of the retreat of the seas, but it is still maintained today. And the general movement of the sea from the east to the west has produced an effect on the surface of the terrestrial mass that is also general, that is, to have steepened all the western coasts of the terrestrial continents and to have at the same time left all of the terrains on the eastern coast with a gentle slope.

As the seas fell and uncovered the highest parts of the continents, these summits, like ventilators that one has just unblocked, began to permit the exhalation of new fires produced in the interior of the Earth by the effervescence of materials that served as fuel for the volcanoes. The domain of the Earth, at the end of this second period of twenty thousand years, was divided between fire and water; equally torn apart and devoured by the fury of these two elements, there was nowhere any security or peace; but happily these ancient scenes, the most horrible in Nature, had no spectators, and it is only after this second period had entirely run its course that one can date the birth of the terrestrial animals. The waters had by then retreated, since the two great continents were united in the north and both inhabited by elephants. The number of volcanoes also had much diminished because their eruptions could not operate except by the conflict of water and fire, and they ceased when the sea had lowered and retreated. One can recall again the aspect that the Earth offered immediately after this second period, that is to say, fifty-five or sixty thousand years after its formation. In all the low areas, deep ponds, rapid currents, and vortices of water; almost continual earthquakes, produced by the collapse of caverns and by the frequent explosions of volcanoes, as much beneath the sea as on land; general and particular storms; billows of smoke and tempests excited by the violent shocks of the ground and of the sea; inundations, floods; deluges brought about by these same commotions; flows of molten glass, of bitumen, and of sulfur ravaging the mountains and entering the plains, poisoning the waters; the Sun almost always obscured not only by watery clouds, but by thick masses of cinders and stones thrown out by the volcanoes, and we thank the Creator not to have made humans witness to these terrifying and terrible scenes, which preceded, and so to speak announced, the birth of sentient and intelligent Nature.

FIFTH EPOCH

When the Elephants and the Other Animals of the South Lived in the North

All that exists today in living Nature could equally have existed since the temperature of the Earth came to be the same. And the northern countries of the globe long enjoyed the same degree of warmth as do today the southern countries; and at the time when the countries of the north enjoyed this temperature, the terrains toward the south were still burning and remained as deserts over a long period of time. It even seems that the memory of this has been conserved by tradition, because the Ancients were persuaded that the lands of the torrid zone used to be uninhabited. They were in fact uninhabitable long after the lands of the north became populated; because, in supposing thirty-five thousand years as the time necessary to cool the Earth and the poles only to the point at which the surface can be touched without fear of being burnt, and twenty or twenty-five thousand years more, as much for the retreat of the seas as for the cooling necessary to allow the existence of beings as sensitive as are terrestrial mammals, one will easily sense that one needs to count a few thousand more years for the cooling of the globe at its equator, as much because of the greater thickness of the Earth there as for its exposure to solar heat, which is considerable at the equator and almost absent beneath the poles.

And all the same these two united causes would not be sufficient to produce such a great difference in time between these two populations. One must consider that the equator received its waters from the atmosphere much later than did the poles, and that by consequence this secondary cause of cooling acted more promptly and more powerfully than the two first causes, the heat of the lands of the north would have been considerably cooled by its receipt of the waters, while the heat of the southern lands persisted and could diminish only by its own loss. And even though one would object to me that the downfall of the waters, whether on the equator, or on the poles, was the result only of cooling to a certain degree of each of these two parts of the globe: it took place in one and the other only when the temperature of the Earth and that of the falling waters was the same, and that consequently this downfall of water did not contribute as much as I say to accelerate the cooling at the pole more than that at the equator. One will be forced to accept that the vapors, and by consequence the waters falling on the equator, possessed more heat through the action of the Sun, and that for this reason they cooled the lands of the torrid zone more slowly, so that I allow at least nine to ten thousand years between the time of the birth of the elephants in the northern countries and the time when they retreated to the more southern lands. Because, the cold came and comes still from above; the continual rains that fell on the polar parts of the globe ceaselessly accelerated the cooling, while no exterior cause contributed to that of the equatorial regions. And this cause, which seems to us so reasonable by the snows of our winters and the hail of our summers, this cold that periodically arrives from the high regions of the air, fell vertically and without interruption on the northern lands, and cooled them more quickly than could be cooled the lands of the equator, upon which these agents of cold—winter, snow, and hail—could neither act nor fall. Besides, we must bring in a very important consideration here on the limits that govern the duration of living Nature. We have established its first possible presence at thirty-five thousand years after the formation of the terrestrial globe, and its last one at ninety-three thousand years from this day, which makes one hundred and thirty-two thousand years for the absolute duration of this beautiful Nature. Here are the most extended limits and the greatest duration that we have given, following our hypotheses as to the life of sensitive Nature. This life could have begun at thirty-five or thirty-six thousand years, because then the globe was cool enough in its polar regions that one could touch it without being burnt, and it can only finish in ninety-three thousand years, when the globe will be colder than ice. But between these so-distant limits, one must admit other, nearer ones: the waters and all the matter that fell from the atmosphere stopped being in a state of boiling only from the moment that one could touch them without being burnt. It was thus only long after this period of thirty-six thousand years that beings endowed with a sensibility comparable to that which we know in them could have been born and survived. Because, if the earth, air, and water suddenly took the degree of heat which would not allow us to be able to touch them without being badly hurt, would there be a single being today capable of withstanding this lethal heat, because it would greatly exceed the vital heat of their own bodies? There could exist then plants, shellfish, and fish of a nature less sensitive to heat, of which the species were destroyed by the cooling in subsequent ages, and these are the ones of which we find the remains and detritus in the coal seams, in the slates, in the shales, and in the beds of clay, just as in the banks of marble and of other calcareous materials. But all the more sensitive species and particularly the terrestrial animals could be born and multiply only in later times, nearer to our own.

And in which country of the north would the first terrestrial animals have been born? Is it not likely that this was in the more elevated regions, because they would have cooled before the others? And is it not equally likely that the elephants and the other animals currently inhabiting the lands of the south were born the first of all, and that they occupied these northern lands for some thousands of years, and long before the birth of the reindeer, which today live in these same lands of the north?

In this time, which is barely more distant from ours than by fifteen thousand years, elephants, rhinoceri, hippopotami, and probably all the species that can multiply today only in the torrid zone, lived and multiplied in the lands of the north, where the temperature was of the same degree, and by consequence suited to their nature. They were there in great number, and they remained there for a long time; the quantity of ivory and of all their other remains that has been discovered, and that one finds still in the northern countries, show us clearly that these were their fatherlands, their birthplace, and certainly the first land that they occupied. But, furthermore, they existed at the same time in the northern countries of Europe, of Asia, and of America. That shows us that these two continents were then contiguous, and that they became separated in subsequent times. I have said that we have in the Cabinet du Roi elephant tusks found in Russia and Siberia, and others that have been found in Canada, near the Ohio River. The large molar teeth of the hippopotamus and the enormous animal of which the species is lost have come to us from Canada, and other very similar ones have come from Tartary and Siberia. One thus cannot doubt that the animals that today inhabit only the southern lands of our continent did not also live in the northern lands of the other continent and at the same time, for the Earth was equally warmed or cooled to the same degree in both. And it is not only in the lands of the north where remains of the animals of the south have been found, but they occur also in the temperate countries, in France, in Germany, in Italy, in England, and so on. We have in this respect authentic monuments, that is to say, the tusks of elephants and other bones of these animals found in several provinces of Europe.

In the preceding times, these same northern lands were covered by the waters of the sea, which by their movement, there produced the same effects as everywhere else: they shaped the hills, they composed the horizontal beds, they deposited the clays and the calcareous matter in the form of sediments. Because, one finds in these lands of the north, as in our countries, shells and the debris of other marine productions buried to quite large depths in the interior of the earth—while it is only so to say within its superficial part, that is to say at a few feet in depth, that one finds the skeletons of elephants, of rhinoceri, and other remains of terrestrial animals.

It even seems that these first terrestrial animals were, like the first marine animals, larger than they are today. We have talked of these enormous square teeth with rounded points, which belonged to an animal larger than an elephant, and of which the species no longer survives. We have pointed out these coiled shells, which can be up to eight feet in diameter and a foot across, and we have similarly seen tusks, teeth, shoulder blades, and femurs of elephants that are larger than those of elephants living today. We have recognized by the direct comparison of molar teeth of the hippopotamus of today with the large teeth that have come to us from Siberia and Canada that the ancient hippopotami to which these large teeth once belonged were at least four times more voluminous than are hippopotami living today. These great bones and these enormous teeth are enduring witnesses of the great force of Nature in these first ages. But, so as to not lose our principal object from view, let us follow the elephants in their progressive march from north to south.

We cannot doubt that after having occupied the northern parts of Russia and Siberia to the sixtieth parallel, where their remains have been found in great quantities, they then reached less northern lands, because these same remains are still found in Poland, in Germany, in England, in France, in Italy. So that, as the lands of the north cooled, these animals sought warmer lands; and, it is clear that all the climes from the north to the equator successively enjoyed a degree of heat agreeable to their nature. Thus, although in human memory the elephant species seems to have occupied only the warmest climes of our continent, that is to say, the lands that extend about twenty degrees on either side of the equator, and that they seem to have been confined there for several centuries, the monuments of their remains, found in all the temperate parts of the same continent, show that they also inhabited, over a similar number of centuries, the different climes of the same continent: at first from the sixtieth to the fiftieth parallel, then from the fiftieth to fortieth, after that from the fortieth to thirtieth, and from the thirtieth to the twentieth; and finally from the twentieth to the equator, and beyond at the same distance. One can even presume that in making investigations in Lapland, in the lands of Europe and Asia that are beyond the sixtieth parallel, one could similarly find there the tusks and bones of elephants and those of other animals of the south, unless one would like to suppose (something that is not unlikely) that the surface of the Earth, being yet more elevated in Siberia than in all the neighboring provinces of the north, these same Siberian lands were the first to be abandoned by the waters, and by consequence the first where terrestrial animals could be established. Whatever it be, it is certain that elephants lived, produced, and multiplied for centuries in this same Siberia and in the north of Russia, that then they reached the lands of the fiftieth to the fortieth parallel, and that they survived there longer than in their country of origin, and for yet longer in the countries of the fortieth to thirtieth parallel and so on, because the progressive cooling of the globe always became slower, as the climes found themselves nearer to the equator, as much due to the greater thickness of the globe as to the greater heat of the Sun.

We have fixed, following our hypothesis, the first possible instant of the beginning of living Nature at thirty-five or thirty-six thousand years from the formation of the globe, because it is only at this instant that one could begin to touch it without being burnt. In giving twenty-five thousand years more to achieve the immense work of the construction of our limestone mountains, for their shaping by salient and reentrant angles, for the lowering of the seas, for the ravages of the volcanoes, and for the drying of the surface of the Earth, we will count only around fifteen thousand years since the time when the Earth, after drying out, and having suffered so many upheavals and changes, at last found itself in a calmer and generally stable state, so that the forces of destruction were not more powerful and more general than those of production. Giving thus an age of fifteen thousand years to living Nature, such as now appears, that is to say, fifteen thousand years of age to the species of terrestrial animals born in the northern lands and still existing in those of the south, we can suppose that the elephants have been confined to the torrid zone for perhaps five thousand years, and that they lived for such a length of time in the climes that today form the temperate zones, and maybe as long in the northern climes, where they originated.

But this steady march that was followed by the largest, the first animals on our continent seems to have been beset by obstacles in others. It is very certain that one has found, and it is very probable that one will find again, the tusks and bones of elephants in Canada, in the lands of Illinois, in Mexico, and in some other parts of North America. But, we have no observation, no monument that shows us the same fact for the lands of southern America. Moreover, the same species of elephant that still lives on the old continent no longer exists on the other. Not only this species, but no others of all the terrestrial animals that now occupy the southern lands of our continent are found in the southern lands of this New World, but it even seems that they existed only in the northern countries of this new continent, and at the same time that they existed in those of our continent. Does not this fact show that the old and the new continent were not then separated in the north, and that their separation took place only after the time when elephants existed in North America, where their species was probably extinguished by the cooling, and about at the time of this separation of the continents—because these animals could not reach the regions of the equator in this continent as they did in the old, as much in Asia as in Africa? Indeed, if one considers the surface of this new continent, one sees that the southern parts near the Isthmus of Panama are occupied by very high mountains: the elephants could not break through these barriers that were impassable to them, because the cold felt at these heights is too great. They were thus not present south of these lands of the isthmus, and would have lived in North America only as long as there was the warmth necessary for their multiplication in those lands. It is the same with all the other animals of the southern parts of our continent; none are found in the southern parts of the other. I have demonstrated this truth by such a great number of examples that one cannot place this into doubt.

The animals, to the contrary, that now inhabit the temperate and cold regions, are equally found in the northern parts of the two continents. They were born there later than the first ones and were conserved there, because their nature does not demand such great warmth. The reindeer and the other animals that can subsist only in the coldest climates have appeared the last, and who knows if over time, when the Earth will become cooler, there will not appear new species of which the temperament will differ from that of the reindeer as much as the nature of the reindeer differs in this respect from that of the elephant? However it is, it is certain that none of the animals that belong and are particular to the southern lands of our continent are found in the southern lands of the other, and that even in the number of animals common to our continent and to that of North America, of which the species are conserved in both, one can hardly cite one that has arrived in South America. This part of the world has thus not been inhabited like all the others nor at the same time; it remained so to speak isolated and separated from the rest of the Earth by the seas and by its high mountains. The first terrestrial animals born in the lands of the north thus could not have been established by communication in this southern continent of America, and could exist only in its northern continent as long as that retained the degree of warmth necessary for their propagation. And, this land of South America, reduced to its own forces, produced only animals more feeble and very much smaller than those that came from the north to inhabit our countries of the south.

I say that the animals that today inhabit the lands of the south of our continent have come there from the north, and I believe I am able to affirm this with all foundation. For, on one side the monuments that we have just revealed demonstrate this, and on the other hand we do not know of any of the great and principal species, currently living in the lands of the south, that did not previously exist in the lands of the north, because one finds there the tusks and bones of elephants, the skeletons of rhinoceri, the teeth of hippopotami, and the monstrous heads of oxen, which are striking in their grandeur, and that it is more than probable that one has also found there the debris of several other less remarkable species. By this means, if one wishes to distinguish in the southern lands of our continent the animals that arrived there from the north from those which that very same land could produce by its own forces, one will recognize all those that are colossal and great in size in Nature have been formed in the lands of the north, and if those of the equator have produced a few animals, these are the inferior species, much smaller than those first ones.

But what must place this production into doubt is that the species that we presume produced here by the inherent forces of the southern lands of our own continent should have resembled the animals of the southern lands of the other continent, which similarly were only produced by the specific force of this isolated land. It is nevertheless just the opposite, as none of the animals of South America sufficiently resemble the animals of the south of our continent to enable us to regard them as the same species. They are for the most part of a form so different, that it is only after long examination that one can suspect them to be the representatives of some of those of our continent. What a difference between the elephant and the tapir, which, however, is the only one of all that can be compared to it, but which is removed already by its shape and prodigiously so in size, as this tapir, this elephant of the New World, has neither trunk nor tusks, and is hardly larger than a donkey. No animal of South America resembles a rhinoceros, none a hippopotamus, none a giraffe; and what a difference between the llama and the giraffe, although it is not as great as that between the tapir and the elephant.

The establishment of living Nature, above all that of the terrestrial animals, thus happened in South America well after its dwelling place had been fixed among the lands of the Earth, and perhaps the difference in time might have been more than four or five thousand years. We have laid out a part of the facts and the reasons that must make one think that the New World, especially in its southern parts, is a land more recently inhabited than was our continent; and that Nature, far from being degenerate through age and decay, was there, to the contrary, born late, and never existed there with the same forces, the same active power as in the northern countries. Because, one cannot doubt, after what we have just stated, that the first and great productions of living beings was made in the high lands of the north, from where they successively passed into the countries of the south in the same form and without having lost anything save for the dimensions of their size. Our elephants and our hippopotami, which to us seem so large, had yet larger ancestors in the times when they inhabited the northern lands, where they left their remains. The cetaceans of today are similarly smaller than they were formerly, though this is perhaps for another reason.

The whales, the rorquals, cachalots, narwhals, and other great cetaceans belong to the northern seas; while one finds, in the temperate and southern seas, only the manatees, dugongs, the porpoises, which all are inferior in size to the former. It seems thus, at first glance, that Nature has operated in a contrary manner and by an inverse succession, since all the largest animals today are found the countries of the south, while all the largest marine animals inhabit only the regions of our pole. And why do these great and almost monstrous species seem to be confined to these cold seas? Why did they not successively extend, like the elephants, into the warmer regions? In a word, why are they not found either in the temperate seas or in those of the south? Because with the exception of some cachalots, which are found frequently enough around the Azores and sometimes are stranded on our coasts, and of which the species seems to be the most vagabond of all the great cetaceans, all the others remain and still have their home in the boreal seas of the two continents. It has been remarked on since people began to fish for, or rather to hunt, these great animals, that they have left the regions where humans went to disturb them. It has been further observed that the first baleen whales, that is to say, those that have been fished for a hundred and fifty and two hundred years, were much larger than those of today: they were up to a hundred feet long; while the ones that are caught today are no more than sixty feet long. One can even explain this difference in size in a reasonably satisfactory manner. Because the baleen whales, like all the other cetaceans, and even most fish, live incomparably longer than any terrestrial animals; and from this, their whole growth also demands a much longer time. Now when people began the fishery of baleen whales, one hundred and fifty or two hundred years ago, one found the oldest ones and those that had completed their growth. They were pursued, hunted by preference, and finally destroyed, and today there remain, in the seas frequented by our fishermen, only those that have not yet attained their full dimensions. Because, as we have said elsewhere, a baleen whale might well live for a thousand years, since a carp can live for more than two hundred.

The continual presence of these great animals in the boreal seas seems to furnish a new proof of the continuity of the continents in the regions of our north, and shows us that this continuity has lasted a long time. For, if these marine animals, which we presume for the moment to have been born at the same time as the elephants, would have found the route open, they would have spread to the seas of the south, if the cooling of the waters had been for them the reverse. And, this would have happened, if they had been born in the times when the sea was still hot. One thus must presume that their existence is subsequent to that of the elephants and the other animals that can exist only in the climate of the south. Moreover, it could also be that the difference in temperature could be, so to speak, unimportant to, or much less felt by, aquatic than terrestrial animals. The cold and the warmth on the surface of the land and of the sea, in truth, follow the pattern of the different climes, and the heat of the interior of the globe is the same in the depths of the sea as it is below the land at the same depth. But, the variations in temperature, which are so great on the surface of the land, are much smaller and almost absent at a few toises of depth below the waters. The insults of the air are not felt there, and the great cetaceans do not suffer them or at least can protect themselves from them. Besides, by the very nature of their organization, they seem to be better provided against the cold rather than against great heat; as although their blood is about as warm as that of quadruped animals, the enormous quantity of lard and oil that covers their body deprives them of the keen sensitivity that other animals possess, protecting them at the same time from all the exterior sensations, and it is to be presumed that they stay where they are because they do not even have the feeling that could guide them to a gentler temperature, nor the idea to find themselves better elsewhere, because instinct is needed to make oneself comfortable, and is needed to decide a change in habitat. There are animals and even people that are so brutish that they prefer to languish in their poor birthplace, rather than the trouble they would need to take to find more comfortable surroundings elsewhere.[1] It is thus very probable that the cachalots, that we see from time to time coming to our coasts from the northern seas, did not undertake this voyage to enjoy a milder temperature, but they are brought there by the shoals of herring, of mackerel, and other small fish that they follow and devour in their thousands.

All these considerations suggest to us that the regions of our north, whether the sea or the land, not only were the first made fertile, but it was also in these same regions that living Nature was raised to its greatest dimensions. And how can one explain this superiority of force and priority of origination given to this region of the north exclusively among other parts of the Earth? Because we see by the example of South America, in the lands where only small animals are found, and in the seas solely the manatee, which is as small by comparison with a whale as the tapir is by comparison with the elephant. We see, I say, by this striking example, that Nature never produced, in the lands of the south, animals comparable in grandeur with the animals of the north; and we similarly see, by a second example drawn from the monuments, that in the southern lands of our continent, the largest animals are those that have come from the north. And of those that were produced in the lands of our south, these are only species greatly inferior to the first ones in grandeur and in force. One can even believe that none were produced in the southern lands of the ancient continent, although some were formed in those of the new; and here are the reasons for this presumption.

All production, all generation, and even all growth, all development, supposes the combination and the reunion of a great quantity of living organic molecules. These molecules, which animate all organized bodies, are successively employed in the nutrition and in the generation of all beings. If the greater part of these beings was suddenly suppressed, one would see new species appear, because these organic molecules, which are indestructible and always active, would reunite to form other organized bodies. But, in being entirely absorbed by the inner molds of beings currently alive, they cannot form new species, at least in the first ranks of Nature, such as those of the great animals. And these great animals have arrived from the north onto the lands of the south. They were there nourished, they reproduced, multiplied, and by consequence absorbed the living molecules, such that they did not leave any superfluous material that could have formed new species. While, by contrast, in the lands of South America, where the great animals of the north could not penetrate, the living organic molecules not being absorbed by any then-existing animal mold, were reunited to form species that did not at all resemble others, and that are all inferior, as much in force as in grandeur, to those of the animals that came from the north.

These two originations, although of different times, took place in the same manner and by the same means; and if the first are superior in all respects to the last, it is that the fertility of the Earth, that is to say, the quantity of living organic matter, was less abundant in the southern climes than those of the north. One can give a reason for this, without searching beyond our own hypothesis. Because all the aqueous, oily, and ductile parts that can enter into the composition of organized beings had fallen with the waters on the northern parts of the globe, much sooner and in much larger quantity than on the southern parts, it is in these aqueous and ductile materials that the living organic molecules began to exercise their power to shape and develop organized bodies. And, as the organic molecules were only produced by heat upon the ductile materials, they were also more abundant in the lands of the north than they could be in the lands of the south, where these same materials were in lesser quantity, it is not surprising that the first, the strongest, and the greatest productions of living Nature were made in these same lands of the north. While, in those of the equator, and particularly in those of South America, where the quantity of these ductile materials was much less, there were formed only inferior species, smaller and more feeble than those of the lands of the north.

But let us return to the principal subject of our epoch. In this same time when the elephants lived in our northern lands, the trees and the plants that today cover our southern lands existed also in these same lands of the north. The monuments seem to demonstrate this, as all the well-known impressions of plants that have been found in our slates and our coals show the character of plants that exist today only in the East Indies or in other parts of the south. One could question me, in spite of the certainty of this fact by the evidence of these proofs, because the trees and the plants cannot travel like the animals, nor by consequence move from north to south. To that I reply: First, this transport did not happen suddenly, but progressively. The species of plants sowed themselves more or more closely toward the lands whose temperatures suited them. And then, these same species, having reached as far as the countries of the equator, would have perished in those of the north, of which they could no longer tolerate the cold. Second, this transport, or rather this successive encroachment of trees, is not even necessary to give reason for the existence of these plants in the southern countries as, in general, the same temperature, that is to say, the same degree of warmth, produces everywhere the same plants without them being transported there. The population of the southern lands by plants is thus even simpler than that by animals.

There remains the spread of man: was this contemporary with that of animals? The major indications and very solid reasons unite here to prove that this happened after all our epochs, and that man is in effect the great and last work of creation. One will not fail to tell us that the analogy seems to demonstrate that the human species has followed the same path and that it dates from the same time as the other species, and that it is even more universally disseminated. And that, if the epoch of its creation is later than that of animals, nothing proves that man was not at least subject to the same laws of Nature, the same alterations, the same changes. We agree that the human species does not essentially differ from other species by its bodily faculties, and that in this regard its fate would have been more or less the same as that of other species. But, can we doubt that we differ prodigiously from the animals by the divine ray that it pleased the sovereign Being to bestow upon us? Do we not see that in man, the matter is led by the spirit? He has thus been able to modify the effects of Nature; he has found a means to resist the intemperance of climate. He created heat, when the cold was destroying him; the discovery and uses of the element of fire, due solely to his intelligence, made him stronger and more robust than any of the animals, and placed him in a state to endure the unhappy effects of cooling. Other arts, that is to say, other traits of his intelligence, provided clothes, arms, and soon he found himself master of the domain of the Earth. These same arts have given him the means to traverse all of its surface, and to become settled everywhere; because, with more or fewer precautions, all climes became for him more or less equal. It is thus not astonishing that, although there is no animal of the south of our continent in the south of the other, man alone, that is to say, his species, is found equally in this isolated land of South America, which seems to have played no part in the first formations of animals, and also in all cold or warm parts of the surface of the Earth. Because, in whatever part and however far one has penetrated since the perfection of the art of navigation, man has found men everywhere. The most ill-favored countries, the most isolated islands, the most distant of continents, have nearly all been found to be inhabited. And, one cannot say that humans, such as those of the Marianas Islands, or those of Tahiti and the other islands located in the middle of the seas and at such great distances from all inhabited lands, are not nevertheless humans of our species, because they can reproduce with us, and that the small differences that we can see in their nature are only slight variations caused by the influence of climate and of nourishment.

Nevertheless if we consider that man, who can so easily guard himself against the cold, cannot by contrast defend himself by any means against too great a heat—that he even suffers greatly in the climes that the animals of the south seek out by preference—one will have one more reason to believe that the creation of man was later than that of the great animals. The sovereign Being did not spread the breath of life in the same instant over all the surface of the Earth; He began by making fertile the seas and then the most elevated lands; and He wished to give all the time necessary for the Earth to solidify to take its shape, to cool, to reveal itself, to dry out, and at last arrive at a state of repose and tranquility, where man could be the intelligent witness, the peaceful admirer of the great spectacle of Nature and of the marvels of creation. Thus we are persuaded, independently of the authority of the sacred texts, that man was created the last, and that he came to take the scepter of the Earth only when it was worthy of his dominion. It nevertheless seems that he first lived, like the terrestrial animals, in the high lands of Asia; and it is in these same lands that the arts of prime necessity were born, and soon after the sciences, which are equally necessary to the exercise of the power of man, and without which he could not form society, nor measure his life, nor have command over animals, nor have use of plants other than by grazing on them. But we will leave to our last epoch the principal facts that bear upon the history of the first humans.

SIXTH EPOCH

When the Separation of Continents Was Made

The time of the separation of continents was certainly later than the times when the elephants lived in the lands of the north, as their species then subsisted equally in America, in Europe, and in Asia. This is shown to us by the monuments, which are the remains of these animals found in the northern parts of the new continent, as in those of the old. But how did it happen that this separation of continents seems to have been made in two places, by two stretches of sea that extend from the northern countries, always becoming wider, as far as the most southern countries? Why do these stretches of sea not occur, by contrast, about parallel to the equator, since the movement of the seas is in general from east to west? Is this not a new proof that the waters originally came from the poles, and that they only progressively reached the regions of the equator? As long as the fall of the waters lasted, and up to the entire purging of the atmosphere, their general movement was directed from the poles to the equator. And, as they came in much greater quantity from the southern pole, they formed vast seas in that hemisphere, which went, diminishing ever more, into the northern hemisphere, up to the polar circle. And, it is by this movement directed from south to north, that the waters sharpened all the terminations of the continents. But, after their complete establishment across the surface of the Earth, where they reached everywhere above two thousand toises, was their movement from poles to equator not combined, before ceasing, with movement from east to west? And when it had completely ceased, did not waters entrained by the sole movement from east to west make scarps of all the western edges of the terrestrial continents, while they successively became lower? And finally was it not after their retreat, when all the continents had appeared, that their contours took on their final form?

We see first that the extent of the lands in the northern hemisphere, taking that from the polar circle to the equator, is so great by comparison with the extent of lands similarly considered in the southern hemisphere, that one could regard the first as the terrestrial hemisphere and the second as the marine hemisphere. Moreover, the distance between the two continents is so small near the regions of our pole, that one can hardly doubt that they were continuous in the times that followed the retreat of the waters. If Europe is today separated from Greenland, that is probably because there was a considerable subsidence between the lands of Greenland and those of Norway and the tip of Scotland, of which the Orcades, the island of Shetland, those of the Faroes, of Iceland, and of Hola, now reveal to us only the summits of submerged terrains. And, if the continent of Asia is no longer contiguous with that of America in the north, this is doubtless a consequence of a similar effect. This first subsidence, which the volcanoes of Iceland seem to demonstrate to us, was not only later than the subsidence of the countries of the equator and the retreat of the seas, but also later by some centuries than the birth of the great terrestrial animals in the northern countries. And, one cannot doubt that the separation of the continents toward the north was not of generally modern date by comparison with the division of these same continents toward the equatorial regions.

We can further infer that not only was Greenland once joined with Norway and Scotland, but also that Canada could have been linked with Spain by the banks of Newfoundland, the Azores, and the other islands and sea mounts that are found in that interval of the seas. They seem to represent to us today the most elevated summits of the lands sunk beneath the waters. This submergence is perhaps yet more modern than that of the continent of Iceland, since its tradition seems to have been conserved. The story of the island of Atlantis reported by Diodore and Plato could only apply to a very large land that extended far to the west of Spain; this land of Atlantis was densely inhabited, governed by powerful kings who commanded several thousand soldiers, and this indicates strongly enough the proximity of America to these Atlantic lands situated between the two continents. We aver nevertheless that the sole thing that is demonstrated here by the facts, is that the two continents were united at the times when elephants existed in the northern countries of the one and the other, and there is, according to me, a much greater probability of this continuity of America with Asia than with Europe. Here are the facts and the observations upon which I have based this opinion.

1. While it is probable that the lands of Greenland were joined to those of America, one cannot be assured of this, because this terrain of Greenland is separated from it first by the Davis Strait, which is not very large, and then by Baffin Bay, which is somewhat larger; and this bay extends to the seventy-eighth parallel, meaning that it is only beyond this that Greenland and America could have been contiguous.

2. Spitzbergen seems to have been a continuation of the lands of the eastern coast of Greenland, and there is a fairly large interval of sea between this coast of Greenland and that of Lapland. Thus, one can scarcely imagine that the elephants of Siberia or of Russia could have reached Greenland. It is the same with their passage by the stretch of land that one can suggest between Norway, Scotland, Iceland, and Greenland, as this interval shows us seas of considerable extent. Besides, these lands, like those of Greenland, are farther to the north than those where elephant bones have been found, as much in Canada as in Siberia. It is thus not probable that it was by this route, now completely obliterated, that these animals could have traveled from one continent to the other.

3. Although the distance from Spain to Canada is far greater than that from Scotland to Greenland, this route seems to me to be the most natural of all, if we are forced to admit the passage of elephants from Europe to America, because this great interval of sea between Spain and the neighboring parts of Canada is considerably shortened by the banks and islands scattered across it; and giving some greater probability to this presumption is the story in tradition of the submergence of Atlantis.

4. One can see, of these three routes, the first two seem impracticable, and the last so long that there is little likelihood that the elephants could pass from Europe to America. At the same time there are very strong reasons that lead me to believe that this communication of elephants, from one continent to another, must have taken place through the northern lands of Asia neighboring America. We have observed that in general all the coasts, all the slopes of the land are steeper toward the seas of the west, which, for this reason, are generally deeper than the seas in the east. We have seen that by contrast all the continents extend as long gentle slopes toward the seas of the east. One can thus presume, with some basis, that the eastern seas beyond and above Kamchatka are not very deep. And, one has already seen that they are littered with a very great number of islands, of which some form terrains of a vast extent: this is an archipelago that extends from Kamchatka for half the distance from Asia to America along the sixtieth parallel, and which seems to touch it beneath the polar circle, by the islands of Anadir, and by the tip of the continent of Asia.

Besides, the voyagers that have frequented both the western coasts of the north of America and the eastern lands from Kamchatka to the north of this part of Asia agree that the natives of these two lands of America and Asia resemble each other so strongly that one can hardly doubt that they were issued one from the other. Not only do they resemble each other in height, by the form of their features, the color of their hair, and the form of their bodies and limbs, yet also by their customs and even by their language. There is thus a very great probability that it is from these lands of Asia that America received its first inhabitants of all species, at least if one does not wish to contend that the elephants and all the other animals, like the vegetation, were created in great numbers in all the climes where the temperature was suitable for them: a bold supposition and more than gratuitous, since it suffices for two individuals or even a single one, that is to say, of one or two inner molds, once made and given the faculty of reproduction: so that, over a certain number of centuries, the Earth becomes inhabited by all the organized beings of which the reproduction does or does not assume the union of the sexes.

On reflecting upon the traditional story of the submergence of Atlantis, it seems to me that the ancient Egyptians, who passed it down to us, had communications and commerce, by the Nile and the Mediterranean, as far as Spain and Mauretania. It is by this communication that they would have been informed of this fact which, however grand and however memorable it is, would not have come to their knowledge if they had not left their country, which was far from where the event took place. It would therefore seem that the Mediterranean, and even the strait that joins it to the ocean, existed before the submergence of Atlantis; nevertheless, the opening of the strait could well have been of the same date. The cause of the submergence suffered by this vast land must have extended to its surroundings. The same convulsions that destroyed it could have toppled the small portion of the mountains that previously closed the strait. The earthquakes which, even in our days, are still felt violently in the surroundings of Lisbon, show us well enough that they are only the latest effects of an ancient and more powerful action, to which one can attribute the foundering of this portion of the mountains.

But what was the Mediterranean before the rupture of this barrier at the edge of the ocean, and of that which closed the Bosphorus at its other extremity toward the Black Sea?

To reply to this question in a satisfactory manner, one needs to reunite in a single panorama Asia, Europe, and Africa, regarding them only as a single continent, and to represent the shape, in relief, of the surface of all of this continent with the courses of its rivers. It is certain that those that flow into the Aral Lake and the Caspian Sea do not provide as much water as these lakes lose by evaporation. It is certain also that the Black Sea receives, in proportion to its extent, much more water by rivers than is received by the Mediterranean, and the Black Sea discharges through the Bosphorus the water that it has in surplus, whereas, by contrast, the Mediterranean, which receives only a small quantity of water by its rivers, draws it from the ocean and from the Black Sea. Thus, in spite of this communication with the ocean, the Mediterranean

Sea and other interior seas should be regarded only as lakes of which the extent has varied, and which are not today what they used to be in other times. The Caspian Sea must once have been much larger, and the Mediterranean smaller, before the opening of the straits of Bosphorus and of Gibraltar. Lake Aral and the Caspian made up only a single great lake, which was the common sink for the Volga, the Jaik, the Sirderoias, the Oxus, and all the waters that could not flow into the ocean. These rivers progressively supplied silts and sands, which today separate the Caspian from the Aral: the volume of water has diminished in these rivers as the mountains where they entrained the sediments have diminished in height. It is thus very probable that this great lake, which is in the center of Asia, was formerly yet larger, and that it communicated with the Black Sea before the rupture of the Bosphorus. Because, in this supposition, which seems to me well founded,[1] the Black Sea, which today receives more water than it can lose by evaporation, used once to be joined with the Caspian, which receives only as much as it loses, and so the surface of these two seas, united, was extensive enough for all of the waters brought by the rivers to be removed by evaporation.

Besides, the Don and the Volga are so close to each other to the north of these two seas that one can hardly doubt that they were combined in the times when the Bosphorus was still closed, not giving to their waters any issue into the Mediterranean: thus, those of the Black Sea and its outliers then extended over all of the low ground that adjoined the Don, the Donjec, and so on, and those of the Caspian Sea covered the neighboring lands of the Volga, to form a lake that was longer than wide and that reunited these two seas. If one compares the present-day extent of Lake Aral, of the Caspian Sea, and of the Black Sea with the extent that we suppose they had when they were continuous, that is to say, before the opening of the Bosphorus, one will be convinced that the surface of these waters then being more than double that of today, evaporation alone would have sufficed to maintain their equilibrium without their overflowing.

This basin, which was then perhaps as large as the Mediterranean today, received and contained the waters of all the rivers of the interior of the continent of Asia which, because of the location of the mountains, could not flow from any side to arrive at the ocean. This great basin was the common sink for the waters of the Danube, of the Don, of the Volga, of the Jaik, of the Sirderoias, and of several other very considerable rivers that flow into them or that fall directly into these interior seas. This basin, sited in the center of the continent, received the waters of those lands of Europe of which the slopes were directed toward the course of the Danube; that is to say, the greater part of Germany, of Moldavia, of Ukraine, and of European Turkey. It similarly received the waters of a large part of the lands of northern Asia, by the Don, the Donets, the Volga, the Jaik, and in the south by the Sirderoias and the Oxus, which presents a very vast extent of land of which all the waters poured into this common sink, while the basin of the Mediterranean then only received those of the Nile, the Rhône, the Po, and of a few other rivers. And so, in comparing the extent of the lands that provide the water to these last rivers, one will clearly recognize that this extent is smaller by half. We are thus well founded in presuming that before the rupture of the Bosphorus and of the Strait of Gibraltar, the Black Sea united with the Caspian Sea and the Aral and formed a basin double in extent of that which remains, and by contrast the Mediterranean at the same time was smaller by half than it is today.

As long as there were the barriers of the Bosphorus and of Gibraltar, the Mediterranean was hence only a lake of moderate extent, where evaporation sufficed for the receipt of waters from the Nile, the Rhône, and the other rivers that belong to it. But, in supposing, as traditions seem to indicate, that the Bosphorus was the first to open, the Mediterranean would from then have been considerably enlarged, by the same amount that the basin above the Black and Caspian seas would have diminished. This great effect is nothing but natural, because the waters of the Black Sea, lying higher than those of the Mediterranean, act continually by their weight and by their movement against the lands that enclose the Bosphorus. They would have undermined them from their base; they would have attacked the weakest places, or perhaps they would have been brought in by some subsidence caused by an earthquake. And, once they had opened this outlet, they would have inundated all the lower ground and caused the most ancient deluge of our continent, because this rupture of the Bosphorus must instantly have produced a great and permanent inundation, which, since this early time, drowned all the lowest terrains of Greece and of adjacent provinces. And, this inundation at the same time extended over the lands that used to adjoin the Mediterranean basin, which from that time was higher by many feet and would have permanently covered the low ground in its neighborhood, even more so on the coast of Africa than that of Europe; because the coasts of Mauretania and of Barbary are very low by comparison with those of Spain, of France, and of Italy all along this sea. Thus, the continent has lost in Africa and in Europe as much land as it gained, so to speak, in Asia by the retreat of the waters between the Black Sea, the Caspian, and the Aral.

Then, there was a second deluge when the gateway at the Strait of Gibraltar was opened. The waters of the ocean would have produced a second augmentation and would have inundated those lands that were not yet submerged. It is perhaps only at this second time that the Adriatic gulf was formed, together with the separation of Sicily and of the other islands. However it happened, it was only after these two great events that the equilibrium of these two interior seas was established and that they took on their dimensions more or less as we see them today.

Besides, the epoch of the separation of these two great continents, and even that of the rupture of the barriers to the ocean and to the Black Sea, seem to have been much older than the date of the deluges conserved in human memory. That of Deucalion is only around fifteen hundred years before the Christian Era, and that of Ogyges eighteen hundred years before. Both were only local inundations, of which the first ravaged Thessaly and the second the lands of Attica. Both were produced only by local and particular causes that were temporary, as were their effects. A few earthquake shocks could have raised the waters of the neighboring seas and made them sweep across the lands, which would have been inundated for a short time without longer submergence. The deluge of Armenia and of Egypt, of which the memories are conserved by the Egyptians and the Hebrews, are older by some five centuries than that of Ogyges, and this is still quite recent by comparison with the events of which we have just spoken, since one counts only around four thousand, one hundred years since this first deluge, and it is very certain that the times when the elephants lived in the lands of the north were well before this modern date, because we are assured by the most ancient books that ivory came from the southern lands. Consequently, we cannot doubt that the elephants have been living in the lands where they are found today for longer than three thousand years. One should therefore consider these three deluges, memorable as they were, as brief inundations that did not at all change the face of the Earth, while the separation of these two continents neighboring Europe could take place only by submerging for good the lands that had joined them. It is the same with the greater part of the lands presently covered by the waters of the Mediterranean. They have been submerged for all time since the gateways were opened at the two extremities of this interior sea to receive the waters of the Black Sea and those of the ocean.

These events, although later than the establishment of terrestrial animals in the countries of the north, perhaps preceded their arrival in the lands of the south, because we have shown that, in the preceding epoch, many centuries passed before the elephants of Siberia could reach Africa or the southern parts of India. We have counted ten thousand years for this type of migration, which took place only by means of the successive and very slow cooling of the different climes from the polar circle to the equator. Thus the separation of the continents, the submergence of the lands that had united them, those of the lands adjacent to the former Mediterranean lake, and finally the separation of the Black Sea from the Caspian and the Aral, although all later than the establishment of these animals in the countries of the north, could well have preceded their populating of the lands of the south, of which the excessive heat then did not permit sensitive creatures to become accustomed to it, or even approach it. The Sun was still the enemy of Nature in these regions, still burning from their own heat, and it became its generator only when this internal heat of the Earth had cooled sufficiently to not offend the sensibility of beings that resemble us. The lands of the torrid zone have been inhabited for perhaps only five thousand years, while one can count at least fifteen thousand years since terrestrial animals became established in the countries of the north.

The high mountains, although situated in the hottest climes, were cooled perhaps as quickly as those of the temperate countries. Because, being higher than those, they formed peaks that were more distant from the mass of the globe. One must thus consider them independently of the general and successive cooling of the Earth from the poles to the equator. There were specific coolings that were more or less rapid in all the mountains and in all the higher ground of the different parts of the globe. And, in the times of its excessive heat, the only places that would be suitable to living Nature were the summits of the mountains and of the other elevated regions, such as those of Siberia and of the high Tartary.

When all the waters were established on the globe, their movement from east to west steepened the western coasts of all the continents throughout all the time that sea level fell. Then, the same movement from east to west directed the waters against the gentle slopes of the eastern lands, and the ocean took possession of their ancient coasts. And, furthermore, it seems to have shaped all the tips of the terrestrial continents, and to have formed the Straits of Magellan and the tip of America, of Ceylon, and the tip of India, of Frobisher Land, and that of Greenland, and so on.

It is at the date of around ten thousand years in the past, counting from today, that I would place the separation of Europe from America, and it is about the same time that England was separated from France, Ireland from England, Sicily from Italy, Sardinia from Corsica, and both from the continent of Africa. It is perhaps also at the same time that the Antilles, Saint-Domingue, and Cuba were separated from the continent of America. All these particular separations were contemporary with or a little later than the great separation of the two continents. Most of them even seem to be necessary consequences of this great division, which, having opened up a large route to the waters of the ocean, would have permitted them to flow over all the low ground, attacking by their movement the less solid parts, to undermine them little by little, and to finally cut them, to separate them from the neighboring continents.

One can attribute the division between Europe and America to the subsidence of the lands that formerly formed Atlantis. And, the separation between Asia and America (if it really exists) supposes a comparable subsidence in the northern seas of the east, but tradition has conserved for us only the memory of the submergence of Ceylon, a land situated near the torrid zone, and consequently too distant to have influenced this separation of the continents toward the north.[2] Inspection of the globe shows us truly that there were greater and more frequent upheavals in the Indian Ocean than in any other part of the world; and that they took place not only through great changes in these countries by the collapse of caverns, earthquakes, and the actions of volcanoes, but also by the continual action of the general movement of the seas which, constantly directed from the east to the west, took over a great extent of land along the former coasts of Asia, and formed the little interior seas of the Kamchatka, of Korea, of China, and so on. It even seems that they also drowned all the low ground that lay to the east of this continent. Because, if one draws a line from the northern extremity of Asia, passing by the point of Kamchatka as far as New Guinea, that is to say, from the polar circle to the equator, one will see that the Marianas Islands and those of the Calanos, which are found in the direction of this line for a length of more than two hundred and fifty leagues, are the remains or rather the former coasts of these vast lands that were invaded by the sea. Then, if one considers the lands from those of Japan to Formosa, from Formosa to the Philippines, from the Philippines to New Guinea, one will be led to think that the continent of Asia was formerly contiguous with that of New Holland, which sharpens and comes to a point toward the south, like all the other great continents.

These upheavals, so many and so evident in the southern seas, the equally evident invasion of the old eastern lands by the waters of this same ocean, show us well enough the prodigious changes that took place in this huge part of the world, above all in the countries neighboring the equator. However, neither one or the other of these great causes could have produced the separation of Asia and America in the north; it would seem by contrast that if the continents were separated rather than being continuous, the subsidence in the south and the inrush of waters onto the lands of the east would have attracted those of the north, and consequently uncovered the land of this region between Asia and America: this consideration confirms the reasons that I have given above for the genuine contiguity of these two continents toward the north in Asia.

After the separation of Europe and America, after the rupture of the straits, the waters ceased to invade the great spaces, and, subsequently, the land gained more from the sea than it lost. Because, independently of the terrains of the interior of Asia, newly abandoned by the waters, such as those that surround the Caspian and the Aral, independently of all the gentle coastal slopes that this last retreat of the waters left uncovered, the great rivers almost all formed islands and new land near their mouths. One knows that the Delta of Egypt, of which the extent is not inconsiderable, is only ground produced by the deposits of the Nile. It is the same with the great island at the mouth of the river Amur, in the eastern sea of Chinese Tartary. In America, the southern part of Louisiana by the Mississippi River, and the eastern region situated at the mouth of the Amazon River, are lands newly formed by the deposits of these great rivers. But we cannot choose a greater example of recent land than that of the vast terrains of Guyana; their aspect recalls to us the idea of brute Nature, and we will show a nuanced portrait of the successive formation of a new land.

Over an extent of more than a hundred and twenty leagues, from the mouth of the Cayenne River to that of the Amazon, the sea, its level the same as the land, has no floor other than mud, or other coasts than a crown of aquatic trees, of mangroves, or *palétuviers*, of which the roots, stems, and curved branches are all soaked in salt water, and appear only as watery coverts that can be penetrated only by canoe, and with axe in hand. This muddy floor extends along a gentle slope for several leagues, beneath the waters of the sea. On the landward side, beyond this large border of mangroves, of which the branches are inclined more toward the water than raised to the sky, forming a fort, which serves as hideaway for foul animals, there extends still-drowned savannahs, planted with latanier palms and strewn with their debris. These lataniers are large trees, of which in truth the foot is still in the water, but of which the head and the raised branches, laden with fruit, invite the birds to perch. Beyond the mangroves and the lataniers, one finds only the soft woods, the "*comons,*" the "pines," which do not grow in the water, but in the muddy ground where the drowned savannahs terminate, and then begin forests of another type. The ground gently rises and marks, so to speak, its elevation by the solidity and hardness of the trees that they produce. Finally, after some leagues in a direct line from the sea, there are small hills of which the somewhat steep sides, and even the summits, are covered with a great thickness of good earth, where everywhere there are trees of all ages, so closely pressed one against the other that their interlaced crowns barely let through the light of the Sun, and maintain beneath their deep shadow a humidity so cold that the traveler is obliged to light a fire to spend the night there. While, at some distance from these somber forests, in the clearings, the excessive heat during the day is still too hot during the night. This vast terrain of the coasts and of the interior of Guyana is only an equally vast forest, in which a few savages have made a few clearings and small shelters to be able to settle there without losing the enjoyment of the heat of the ground and of the light of the day.

The great thickness of vegetable earth, which is found as far as the summits of the hills, demonstrates the recent formation of all of this country. This is indeed the case to the extent that on top of one of the hills called *Gabrielle*, one can see a little lake with the Cayman crocodiles, which the sea has left, at five or six leagues in distance and six or seven hundred feet in height above its level. Nowhere does one find limestone rock. All the lime necessary for building in Cayenne is transported from France; that which is called "pierre à ravets" is not at all stone, but a volcanic lava, with holes like the scoria of forges; this lava is seen as scattered blocks or as irregular fragments in some mountains where are seen the mouths of old volcanoes, which are now extinct, because the sea has retreated and become more distant from the foot of these mountains. Everything thus combines to prove that it has not been long that the waters have abandoned these hills, and even less time since they left the plains and low ground exposed; because these have almost all been formed by deposition from running water. The rivers, the streams, the creeks are so close one to another and at the same time so large, so widened, so rapid in the rainy season, that they unceasingly entrain immense amounts of silt, which are deposited over all the low ground and at the bottom of the sea as muddy sediments,[3] Thus this new land grows, century by century, while being uninhabited, for one must count as nothing the small number of people that one encounters there. They are still, as much in moral as in physical attributes, in a state of pure nature; neither clothes, nor religion, nor society other than among a few far-dispersed families, perhaps in the number of three or four hundred huts, in a land the extent of which is four times greater than that of France.

These people, like the ground that they inhabit, seem to be the newest in the universe. They arrived there from higher regions and at times that were later than the establishment of the human species in the high countries of Mexico, Peru, and Chile. Because, in assuming the first men were in Asia, they would have passed by the same route as the elephants and on arrival would have spread across the lands of North America and of Mexico. They would have then easily crossed the high ground beyond the isthmus, and would have established themselves in that of Peru, and finally they would have penetrated as far as the most deep-lying countries of South America. But is it not singular that it is in some of these last countries that there exist to this day giants of the human species, while one sees there only pygmies among the types of animals? Because it cannot be doubted that there can be encountered in South America men in great number that are all larger, more squarely built, broader, and stronger than are the other men of the Earth. The races of giants, formerly so common in Asia, no longer exist there. Why are they found in America today? Could we not believe that a few giants, like the elephants, passed from Asia to America, where finding themselves, so to say, alone, their race was preserved on this deserted continent, while it was entirely destroyed by the number of other men in populated countries? One circumstance seems to me to have competed in maintaining this ancient race of giants in the continent of the New World. These are the high mountains that divide it in two along all its length and in all climes. And, one knows that in general inhabitants of the mountains are larger and stronger than those of the valleys or the plains. Supposing, thus, a few pairs of giants that passed from Asia to America, where they found liberty, tranquility, peace, or other advantages that they perhaps did not have at home, would not they have chosen in the lands of their new domain those that suited them the best, as much for the warmth as for the salubriousness of the air and the waters? They would have settled their living quarters at a modest height in the mountains; they would have stopped in the climate most favorable for their multiplication; and as they could have had little occasion for mésalliance, because all the neighboring lands were deserted, or at least all also newly peopled by a small number of people much inferior in strength. Their gigantic race was propagated without obstacle and almost without mixing. It lasted and existed to this day, while for a number of centuries it had been destroyed in the places of its origin in Asia,[4] by the very great and older population of this part of the Earth.

But as people multiplied in the lands that are now hot and temperate, their number diminished in the lands that became too cold. The north of Greenland, Lapland, Spitzbergen, Novaya Zemlya, the land of the Samoiedes, and also a part of those lands that neighbor the glacial sea as far as the extremity of Asia to the north of Kamchatka, are now deserted or rather depopulated since fairly recent time. One can see even on Russian maps, that from the mouths of the Olenek, Lena, and Jana rivers, under the seventy-third and seventy-fourth parallel, the route all along the coast of this glacial sea as far as the land of the Tschutschis, was once much visited, and that now it is impracticable or at least so difficult that it was abandoned. These same maps show us that three ships left, in 1647, the common mouth of the Kolyma and Olomon rivers, under the seventy-second parallel. A single one passed the cape of the land of the Tschutschis under the seventy-fifth parallel, and alone arrived, say these same maps, at the Anadir islands, neighboring America under the polar circle. But, as much as I am persuaded of the truth of the first facts, so I doubt that of the last one; because this same map that shows by a suite of points the route of this Russian ship around the land of the Tschutschis shows at the same time, written out in full, that the extent of this land is unknown. And, while one would in 1648 have traversed this sea and visited this point in Asia, it is certain that since this time the Russians, although very interested in this piece of navigation to arrive at Kamchatka and from there to Japan and China, have completely abandoned it. But, perhaps they also kept for themselves the knowledge of this route around this land of the Tschutschis that forms the most northern and most projecting extremity of the continent of Asia.

However it came to be, all the northern regions beyond the seventy-sixth parallel from the north of Norway as far as the extremity of Asia are now denuded of inhabitants, with the exception of a few unfortunates whom the Danes and the Russians have settled for fishing, and who alone maintain a remainder of the population and some commerce in this frozen land. The lands of the north, formerly warm enough to allow elephants and hippopotami to multiply, being already cooled to the point of supporting only white bears and reindeer, will be, in a few thousand years, entirely denuded and deserted solely by the effects of cooling. There are even very strong reasons that lead me to believe that the region of our pole, that is as yet unknown, will never be known. Because this glacial cooling seems to me to have taken possession of the pole as far as a distance of seven or eight degrees, and it is more than probable that all of this polar shore, formerly sea or land, is today only ice. And if this presumption has foundation, the circumference and the extent of this ice, far from diminishing, can only increase with the cooling of the Earth.

And if we consider what takes place in the high mountains, even in our climes, we will find there a new proof demonstrating the reality of this cooling, and at the same time draw a comparison that seems to me to be striking. One finds above the Alps, over a distance of more than sixty leagues by twenty, and even thirty, in width in certain places, from the mountains of the Savoy and the canton of Berne as far as those of the Tyrol, an immense extent, almost continuous, of valleys, of plains, and of hills of ice, mostly unmixed with any other material and almost all permanent, and which never completely melt. These great shores of ice, far from diminishing their extent, are growing and extending more and more. They are gaining ground on neighboring and lower terrain; this fact is shown by the crowns of large trees, and even by the tip of a clock tower, which are enveloped in these masses of ice, and which appear only in a few very hot summers, as the ice diminishes by a few feet in height. But, the interior mass, which in some places is a hundred toises thick, has not melted in human memory.[5] It is thus evident that these forests and this clock tower, now buried in this thick and permanent ice, were previously situated on uncovered ground, inhabited, and by consequence less frigid than they are today. It is similarly certain that this successive augmentation of ice cannot be attributed to an increase in water vapor, since all the summits of the mountains that overlook these glaciers are not at all elevated, but, to the contrary, have been lowered with time and with the fall of an infinity of rocks and of masses of debris, which have rolled, either to the bottom of the glaciers, or into the smaller valleys. From then on, the enlargement of these ice terrains is already, and will in future be, the most palpable proof of the progressive cooling of the Earth, of which it is easier to detect the stages in these more advanced parts of the globe than anywhere else. If one thus continues to observe the progress of these permanent glaciers of the Alps, one will know in a few centuries how many years it will take for the glacial cold to possess the land that is now inhabited, and from that one will be able to judge whether I have estimated too much or too little time for the cooling of the globe.

Now, if we carry that idea to the region of the pole, we will easily persuade ourselves that not only is it entirely covered in ice, but even that the circumference and the extent of this ice augments from century to century, and that it will continue to augment with the cooling of the globe. The lands of Spitzbergen, some ten degrees from the pole, are almost entirely ice covered, even in summer: and, through the new attempts that have been made to approach the pole more closely, it seems that only ice has been found, which I consider as extensions of the great glacier that covers the entirety of that region, from the pole for up to seven or eight degrees in distance. The immense glaciers recognized by Captain Phipps at the eightieth and eighty-first parallel, and that everywhere prevented him from further advance, seem to prove the truth of this important fact. Because, one cannot presume that there are springs and rivers of fresh water beneath the pole that can produce and bring forward this ice, because these rivers will be frozen in all seasons. It thus seems that the ice that stopped this intrepid navigator from penetrating past the eighty-second parallel, over a length of more than twenty-four degrees in longitude, it seems, I say, that this continuous ice forms a part of the circumference of this immense glacier of our pole, produced by the progressive refrigeration of the globe. And if one wishes to infer the surface of this zone of ice from the pole as far as eighty-two degrees of latitude, one will see that it covers more than one hundred and thirty thousand square leagues. And, consequently, here is already one two-hundredth part of the globe invaded by refrigeration and destroyed for living Nature. And as the cold is greater in the regions of the southern pole, one must presume that the invasion of ice there is yet greater, since one can encounter some ice on southern coasts from the forty-seventh parallel. But, to consider here only our northern hemisphere, of which we presume that the ice has already invaded a hundredth part, that is to say, all the surface of the portion of the sphere that extends from the pole as far as eight degrees or two hundred leagues of distance, one can easily feel that it is possible to determine the time when this ice started to establish itself at the point of the pole, and subsequently the time of the successive progression of its invasion as far as two hundred leagues. One could deduce from that the progression that is to come, and know in advance what will be the duration of living Nature in all of the climes as far as that of the equator. For example, if we suppose that it was a thousand years ago that the permanent ice had begun to be established under the very point of the pole, and that in the duration of this thousand years, the ice has extended around this point as far as two hundred leagues, which makes a hundredth part of the surface of the hemisphere from the pole to the equator, one can presume that there will still pass ninety-nine thousand years before it can invade all of this extent, supposing a uniform progression of the glacial cold as that of the cooling of the globe. This agrees well enough with the duration of ninety-three thousand years that we have given to living Nature, commencing from this day, and that we have deduced just from the law of cooling. However it is, it is certain that ice is present on all sides at eight degrees from the pole as a barrier and insurmountable obstacle. Because, Captain Phipps has sailed more than a fifteenth part of the circumference in the northeast; and before him, Baffin and Smith have recognized as much in the northwest, and everywhere they found only ice. I am thus persuaded that, if a few other equally courageous navigators undertake to survey the rest of this circumference, they will find themselves barred everywhere by ice that they can neither penetrate nor breach; and that consequently this region of the pole is entirely and forever lost to us. The continual fog that covers these climes, and which is only frozen snow in the air, stops, like all the other vapors, against the walls of these coasts of ice; it forms there new layers and more ice, which augment incessantly and always extend farther and farther, as the globe will cool further.

As for the rest, the surface of the northern hemisphere shows much more land than does that of the southern hemisphere; this difference itself suffices, independently of the other causes mentioned previously, for this latter hemisphere to be colder than the former. Thus, one finds ice from the forty-seventh or fiftieth parallel in the southern seas, while one only encounters it twenty degrees further in the northern hemisphere. One sees, besides, that beneath our polar circle there is half as much more land than water, while all is sea beneath the Antarctic Circle. One can see that between our polar circle and the Tropic of Cancer there is more than two-thirds of land for a third of sea, while between the Antarctic polar circle and the Tropic of Capricorn, there is perhaps fifteen times more sea than land. This southern hemisphere has thus been always, as it still is today, much more watery and much colder than ours, and it is not an illusion that after the fiftieth parallel one never finds there happy and temperate lands. It is thus almost certain that ice has invaded over a great extent beneath the Antarctic pole, and that its circumference extends perhaps much farther than that of the ice of the Arctic pole. These immense glaciers of the two poles, produced by cooling, move like the glaciers of the Alps, always augmenting. Posterity will not be slow to know this, and we believe ourselves to have a basis to presume, following our theory and following the facts that we have shown, to which we must add that of the permanent ice that has formed over the past centuries against the eastern coast of Greenland. One can further add to this the growth of the ice near Novaya Zemlya in the Strait of Waeigats, where passage has become more difficult and almost impracticable. And, finally, the impossibility now of traversing the glacial sea to the north of Asia: because despite what the Russians have said of this,[6] it is very doubtful that the coasts of this sea that are the farthest to the north have been reached or that they have gone around the northernmost point of Asia.

So we are here, as I have proposed, descended from the summit of the scale of time as far as the centuries that are close enough to ours. We have passed the chaos and the light, the incandescence of the globe and its first cooling, and this period of time has been twenty-five thousand years. The second stage of cooling permitted the downfall of the waters and produced the cleansing of the atmosphere since twenty-five to thirty-five thousand years ago. In the third epoch the universal sea was established, and the production of the first shellfish and the first plants, the construction of the Earth's surface as horizontal beds, was the work of fifteen or twenty more thousands of years. At the end of the third epoch and at the beginning of the fourth, the waters retreated, and the currents of the sea cut our valleys, and subterranean fires began to ravage the Earth by their explosions. All these last movements lasted ten thousand more years, and in total sum these great events, these operations, and these constructions suppose at least a succession of sixty thousand years. After which, Nature in its first moment of repose made its most noble productions: the fifth epoch presented us with the birth of the terrestrial animals. It is true that this repose was not absolute, the Earth was not yet altogether tranquil, since it was only after the birth of these first terrestrial animals that the separation of the continents was made, and there took place the great changes that I have just described in this sixth epoch.

Furthermore, I have done what I could to apportion in each of these periods the duration of time to the grandeur of the works. I have tried, following my hypotheses, to trace the successive tableaux of the great revolutions of Nature, without nevertheless having pretended to grasp it at its origin, and still less to have embraced it to its full extent. And contested as my hypotheses are and as imperfect a sketch of Nature as my tableaux are, I am convinced that all who wish to examine this attempt in good faith and to compare it with the model will find enough resemblance to at least satisfy their eyes and to focus their ideas on the greatest objects of natural philosophy.

SEVENTH AND LAST EPOCH

When the Power of Man Has Assisted That of Nature

The first men, witnesses of convulsive movements of the Earth, then still recent and very frequent, having only mountains as refuge against inundations, often chased from these same refuges by the fires of volcanoes, trembling on an Earth that was trembling under their feet, naked in spirit and body, exposed to the curses of all the elements, victims of the fury of ferocious animals, of which they could not avoid being the prey; all equally penetrated by a common feeling of baleful terror, all equally driven by necessity, did they not all very quickly seek to unite, first to defend themselves in numbers, then to help each other and work in concert to make dwellings and weapons? They started by sharpening hard pebbles into the form of axes, these jades, these *lightning stones*, that they thought had fallen from the clouds and formed by thunder, and which nevertheless are only the first monuments of the art of man in a state of pure nature. He would soon draw fire from these same pebbles in striking them one against the other; he would have seized the flames of volcanoes, or profited from the fire of their burning lavas, to communicate it, in order to make daylight in the forests and the undergrowth. Because, with the help of this powerful element, he cleansed, made healthy, purified the terrains that he wished to inhabit. With the stone axe he split and cut the trees, fashioned the wood, shaped his weapons and his most essential tools; and after arming themselves with clubs and other heavy and defensive arms, did not these first men find the means to make weapons lighter to strike at a distance? A nerve, an animal tendon, threads of aloes, or the supple bark of a woody plant would have served them as cord to unite the two extremities of an elastic branch out of which they made their bow. They sharpened other little pebbles to arm the arrow; soon they would have had rope, rafts, canoes, and would have thus maintained themselves or formed small nations formed of a few families, or rather of relatives descended from the same family, as we still see today among the savages who wish to stay savages, and who can do this, in those places where the free space is not lacking, nor game, fish, and fruit. But among all those where space is confined by water or restricted by high mountains, these small nations become more numerous, or are forced to divide up the land between them, and it is from this moment that the Earth became the domain of man. He took possession by his work of cultivation, and attachment to a fatherland very quickly followed these first acts of possession; individual interests being part of national interest, order, police, and laws soon followed, and society assumed its solidity and its force.

Nevertheless, these men, profoundly affected by the calamities of their first condition, and still having before their eyes the ravages of floods, the fires of volcanoes, the chasms opened by the shaking of the earth, conserved a durable and almost eternal memory of these misfortunes of the world: the idea that one must perish by a universal deluge or by a general conflagration; the respect for certain mountains[1] where they were saved from inundations; the horror of other mountains that launched fires more terrible than those of thunder; the vision of these combats of the Earth against the heavens, founded on the fable of the Titans and of their assaults upon the Gods; the idea of the real existence of a malevolent Being, the fear and the superstition that are its first product. All of these feelings founded upon terror have from then on taken possession forever of the heart and mind of man; it is barely reassured today by the passage of time, by the calm that followed these centuries of storms, and ultimately by knowledge of the actions and operations of Nature: a knowledge that could be acquired only after the establishment of some great society in these peaceable lands.

It was not in Africa, nor in the lands of Asia that are farthest to the south, that the great societies could first form. These countries were still burning and deserted. It was not in America, which is evidently only, with the exceptions of its mountain chains, a new land. It is not even in Europe, which only much later received the light from the East, where the first civilized humans were established, since before the foundation of Rome, the happiest countries of this part of the world, such as Italy, France, and Germany, were still only peopled by men who were half-savages. Read Tacitus, on the customs of the Germans: it is the picture of those of the Huron, or rather the habits of the entire human species on leaving the state of Nature. It is thus in the northern countries of Asia that the stem of human knowledge grew; and it was upon the trunk of the tree of science that was raised the throne of man's power. The more he knew, the more he could do, but the more he was able. But also, the less he did, the less he knew. All this assumes active peoples in a happy climate, under a pure sky for observing, upon fertile earth for cultivation, in a privileged country, sheltered from floods, distant from volcanoes, higher, and consequently temperate for longer than the others. And all these conditions, all these circumstances, were united in the center of the continent of Asia, from the fortieth degree of latitude to the fifty-fifth. The rivers that bring their waters into the North Sea, the eastern ocean, into the seas of the south and into the Caspian, all depart from this high ground, which today makes part of southern Siberia and Tartary. It was thus in this terrain that was higher, more solid than the others, because it served them as center and because it was distant by nearly five hundred leagues from all the oceans. It was in this privileged country that arose the first people worthy to bear this name, worthy of all our respect, as creator of the sciences, the arts, and all the useful institutions. This truth is equally revealed to us by the monuments of natural history and by the almost inconceivable progress of ancient astronomy. How were some men, so new, able to find the lunisolar period of six hundred years?[2] I confine myself to this single fact, although one could cite many others as marvelous and as constant. They thus knew as much astronomy as *Dominique Cassini* knew in our days, who was the first to demonstrate the reality and exactitude of this period of six hundred years, knowledge of which neither the Chaldeans, nor the Egyptians, nor the Greeks discovered. It is knowledge that supposes that of the precise movement of the Moon and the Earth, and that demands a great perfection in the instruments needed to observe them: a knowledge that could be acquired only after having acquired all, that was only founded upon a long succession of researches, of studies, and of astronomical works, and presumes at least two or three thousand years of cultivating the human mind to attain this.

These first people were very content, because they became very knowledgeable. They enjoyed several centuries of peace, of repose, of the leisure necessary for this culture of mind upon which depend the fruits of all the other cultures. For, to detect this period of six hundred years, it needed at least twelve hundred years of observations. To be assured of it as certain fact, more than double this time is needed. Here already is three thousand years of astronomical studies, and this should not astonish us, since it needed the same duration for astronomers, counting from the Chaldeans up to us, to recognize this period. And were not these first three thousand years of observations necessarily preceded by several centuries when the science was not yet born? Are six thousand years, counting from today, sufficient to ascend to the most noble epoch in the history of man, and even to follow it in the initial progress that he has made in the arts and in the sciences?

But unfortunately they have been lost, these high and beautiful sciences, they have come to us only as fragments too shapeless to serve us otherwise than as knowledge of their former existence. The invention of the formula that the Brahmins used to calculate the eclipses needs as much science as the construction of our ephemerides, and yet these same Brahmins did not have the least idea of the composition of the universe. They were quite wrong concerning the movement, the size, and the position of the planets, they calculated the eclipses without knowing the theory, guided like machines by an array founded on wise formulae that they did not understand, and that probably their ancestors did not at all invent, because they perfected nothing and did not transmit the least ray of science to their descendants. In their hands these formulae were only practical methods, but they presumed a deep knowledge of which they did not possess the elements, and of which they did not conserve even the least vestiges, and which consequently never belonged to them. These methods could thus only have come from this ancient learned people who had reduced to formulae the movements of the stars, and who by a long suite of observations had arrived not only at the prediction of the eclipses, but at the much more difficult knowledge of the period of six hundred years and of all the astronomical facts that this knowledge demands and necessarily presumes.

I believe that I have a basis to say that the Brahmins did not create these learned formulae, because all of their ideas of physics are contrary to the theory upon which these formulae depend, and if they had understood this theory even in the times when they received its results, they would have retained the science and would not have found themselves reduced today to the greatest ignorance, and given to the most ridiculous presumptions concerning the system of the world—because they believe that the Earth is immobile and rests upon the peak of a mountain of gold; they think that the Moon is eclipsed by aerial dragons, that the planets are smaller than the Moon, and so on. It is thus evident that they never possessed the first elements of astronomical theory, nor even the least knowledge of the principles which suppose the methods that they use. But, I must refer to the excellent work that M. Bailly has just published on ancient astronomy, in which he discusses in depth all that relates to the origin and progress of this science. One will see that his ideas accord with mine, and moreover he has treated this important subject with an inspired wisdom and profound erudition that deserves the praise of all who are interested in the progress of science.

The Chinese, a little more enlightened than the Brahmins, calculated the eclipses roughly enough and have calculated them similarly for two or three thousand years. Because they have never perfected anything, they have never invented anything. Science was thus not born in China any more than in India. Although also neighboring with the Indians, the first knowledgeable people, the Chinese seem not to have drawn anything from them; they do not even have the astronomical formulae of which the Brahmins have conserved the use, and which are nevertheless the first and great monuments of the knowledge and well-being of peoples. It also seems that neither the Chaldeans, the Persians, the Egyptians, nor the Greeks have received anything from these first enlightened people. Because, in these countries of the Levant, the new astronomy is due only to the determined assiduity of the Chaldean observers, and then to the works of the Greeks,[3] which one can date only from the time of the founding of the School of Alexandria. Nevertheless, this science was still very imperfect after two thousand years of the new culture and even after our recent centuries. It thus appears to me certain that these first people, who had invented and cultivated astronomy so happily and for so long, left nothing of this but some fragments of a few results that could be retained in memory, such as the period of six hundred years that the historian Joseph passed down to us without understanding it.

The loss of the sciences, this first wound made upon humanity by the axe of barbarism, was without doubt the effect of an unfortunate revolution, which would have destroyed, perhaps in a few years, the work and the works of several centuries. Because, we cannot doubt that these first peoples, at first as powerful as knowledgeable, did not for a long time maintain themselves in their splendor, because they made such great progress in the sciences, and consequently in all the arts demanded by their study. But it altogether seems that, when the lands to the north of this happy country cooled too much, the men who lived there, still ignorant, wild, and barbaric, would have streamed toward the same country that was rich, fertile, and cultivated in the arts. It is even somewhat astonishing that they seized it and that they there destroyed not only the origins, but even the memory of all science, with the result that perhaps thirty centuries of ignorance followed the thirty centuries of light that had preceded them. Of all these beautiful and first fruits of the human mind, there remain only the dregs. Religious metaphysics cannot be understood, have no need of study, and cannot be altered nor lost except by lack of memory, which can never be lacking once it is struck with wonder. So this metaphysics spread out from this first center of the sciences to all parts of the world. The idols of Calicut are seen to be the same as those of Seleginskoi. The pilgrimages to the Grand Lama, established at more than two thousand leagues of distance; the idea of metempsychosis carried yet farther, adapted as an article of faith by the Indians, the Ethiopians, the Atlanteans; these same ideas, disfigured, were received by the Chinese, the Persians, the Greeks, to arrive as far as to us; all seems to show us that the first stock and common stem of human knowledge belongs to this land in high Asia, and that these sterile or degenerate parts of the noble branches of this ancient stock extended into all parts of the Earth among civilized peoples.

And what can we say about these centuries of barbarism, which passed in pure loss for us? They are buried forever in a profound night: man was then plunged back into the shadows of ignorance, and so to say stopped being man. Because coarseness, followed by the forgetting of duty, started by relaxing the ties of society, and barbarism managed to break them; laws derided or done away with, customs degenerated into savage habits, the love of humanity, although engraved in holy letters, finally wiped from hearts. Finally man, without education, without morals, reduced to leading a solitary and wild life, only offers, in place of an elevated nature, that of a being degraded below the level of an animal.

Nevertheless, after the loss of the sciences, the useful arts to which they had given birth were retained: cultivation of the land, which became more necessary as men became more numerous, more crowded; all of the practices demanded by this same culture, all the arts implied by the construction of edifices, the fabrication of idols and of weapons, the weaving of fabrics, and so on, survived after science; they spread out from place to place, and were perfected more and more. They followed the paths of the great populations: the ancient empire of China arose the first, and almost at the same time those of the Atlanteans in Africa; those of the continent of Asia, those of Egypt, of Ethiopia were successively established, and finally that of Rome, to which our Europe owes its civil existence. It is thus only since around thirty centuries that the power of man has been united with that of Nature, and has extended over the greater part of the Earth; the treasures of its fertility were until then buried, and man placed them in bright daylight. Its other riches, still more profoundly buried, could not hide from his researches, and have become the price of his work. Everywhere, where he conducted himself with wisdom, he followed the lessons of Nature, profited from its examples, employed its means, and chose in its immensity all the objects that could service him or please him. By his intelligence, the animals were tamed, subjugated, broken in, reduced to obey him forever. By his works, the swamps were drained, the rivers contained, their cataracts smoothed, the forests cleared, the land cultivated. By his reflection, time was counted, space was measured, the celestial movements recognized, calculated, represented, the heavens and Earth compared, the universe enlarged, and the Creator respectfully adored. Through his arts derived from science, the seas were traversed, mountains crossed, peoples brought closer, a new world discovered, a thousand other isolated lands became his domain. Finally, the entire face of the Earth today carries the imprint of the power of man, which, though subordinate to that of Nature, often created more than did she, or at least marvelously assisted, so it is with the help of our hands that she developed in all her extent, and that she arrived by degrees to the point of perfection and magnificence that we see today.

Compare in effect brute Nature with Nature cultivated: compare the small wild nations of America with our great civilized peoples; compare even those of Africa, which are only half civilized. See at the same time the state of the lands that these nations live in: you will easily determine the small worth of these men by the slight impression that their hands have made on the soil. Either stupid, or lazy, these half-brute men, these unpoliced nations, large or small, only weigh down the globe without comforting the Earth, starve it without nourishing it, destroy without building, using all and renewing nothing. Nevertheless, the most despicable condition of the human species is not that of the savage, but that of those nations that are a quarter policed, which have always been the real curse of human nature, and which civilized peoples still have trouble to contain today. They have, as we have said, ravaged the first happy land, they tore out the seeds of contentment and destroyed the fruits of science. And with how many other invasions was this first incursion of the barbarians followed! It is these same countries of the north, where once were found all of the good of the human species, and then subsequently there arrived all of its faults. How many times have waves of animals with human faces been seen, always coming from the north, ravaging the lands of the south? Cast your eyes on the annals of all the peoples, you will count there twenty centuries of desolation for a few years of peace and repose.

It took six hundred centuries for Nature to construct her great works, to cool the Earth, to shape its surface and to arrive at a tranquil state. How many centuries will be needed for men to arrive at the same point and cease to trouble, to agitate, and to destroy themselves? When will they recognize that the peaceful working of the lands of their fatherland suffices for their happiness? When will they be wise enough to reduce their pretensions, to renounce their imagined dominance, relinquish their foreign possessions, often ruinous or at least more burden than use? The empire of Spain is as extensive as that of France in Europe, and ten times greater in America—is it ten times more powerful? Would it be even so much if this proud and great nation was limited to draw from its own happy land all the goods that it could provide for itself? The English, these people so judicious, so profoundly thoughtful, did they not make a great mistake in extending too far the limits of their colonies? The ancients seem to me to have had saner ideas about these matters; they planned emigrations only when their population became too great, and when their lands and their commerce no longer sufficed for their needs. The invasions of the barbarians, which one regards with horror, did they not have causes still more pressing as they found themselves too tightly pressed in lands that were thankless, cold, and denuded, and at the same time neighboring other lands that were cultivated, fertile, and covered with all the goods that they lacked?

But also how much did these horrific conquests cost in blood, how much unhappiness, how many losses accompanied them and followed them?

We will not long dwell on the sad spectacle of these revolutions of death and devastation, all produced by ignorance. Let us hope that the equilibrium, however imperfect, which is now present between the powers of the civilized peoples will be maintained and can even become more stable as men sense better their true interests, that they will recognize the value of peace and tranquil contentment, that they will make these the sole object of their ambition, that the princes will scorn the false glory of conquerors and disdain the small vanity of those who, to play a role, provoke them into great movements.

Let us suppose thus a world at peace, and see more closely how much the power of man could influence that of Nature. Nothing seems more difficult, not to say impossible, than to oppose the progressive cooling of the Earth and to raise the temperature of a climate; however, man can do this and has done this.

Paris and Quebec are about at the same latitude and at the same elevation on the globe; Paris would thus be as cold as Quebec, if France and all the countries that neighbor it were as bereft of people, as covered with forest, as bathed in waters as are the lands that neighbor Canada. To cleanse, to reclaim and people a country, is to provide it with warmth for many thousands of years, and this forestalls the only reasonable objection that one could make against my opinion, or to put it better, against the real fact of the cooling of the Earth.

According to your system, people tell me, all the Earth should be colder than it was two thousand years ago; yet, tradition seems to prove the opposite. The Gauls and Germania nourished elks, lynx, bears, and other animals that have since retreated into northern countries. This progression is very different from that which you might presume for them, from north to south. Furthermore history tells us that every year the river Seine was generally ice covered for part of winter. Do not these facts seem directly opposed to the suggestion of the progressive cooling of the globe? They would be, I admit, if France and Germany were like Gaul and Germania: if one had not cut down the forests, drained the swamps, contained the torrents, directed the rivers, and cleared all the lands that were overly covered and charged with the debris of their production. But does one not need to consider that the loss of the heat of the globe took place insensibly; that it needed seventy-six thousand years to cool to the level of the present temperature, and that in another seventy-six thousand years it will not yet be cold enough for the particular warmth of living Nature to be obliterated? Does one then not need to compare this cooling, which is so slow, with the prompt and sudden cold that comes to us from the regions of the air? Recall that there is nevertheless only one thirty-second of difference between the greatest heat of our summers and the greatest cold of our winters; and one will already sense that exterior causes have a much greater influence than the interior cause on the temperature of each clime, and that in all those where the cold of the overlying air is attracted by humidity or pushed by the winds, which bring it down to the surface of the Earth, the effects of these particular causes greatly prevail over the product of the general cause. We can give an example that will leave no doubt on this matter, and that forestalls at the same time any objection of this type.

In the immense extent of the lands of Guyana, which are only thick forests where the Sun can barely penetrate, where the widespread waters occupy large areas, where the closely spaced rivers are neither contained nor directed, where it rains continually for eight months of the year, only over the last century has a very small region of these vast forests around Cayenne started to be cleared. And, already, the difference in temperature in this small area of cleared terrain is so noticeable that one suffers there from too great a heat, even during the night, while over all the rest of the tree-covered ground it is cold enough at night to force one to light a fire. It is the same with the quantity and continuity of the rains. They cease sooner and commence later at Cayenne than in the interior of the country; they are also less abundant and less continuous. There are four months of absolute drought at Cayenne, while in the interior of the country, the dry season lasts only three months, and also it rains every day as a violent enough storm, that one calls the *grain de midi*, because it is toward the middle of the day that this storm begins. Furthermore, there is hardly ever any thunder at Cayenne, while thunder is violent and very frequent in the interior of the country, where the clouds are black, thick, and very low. These facts, which are certain, do not they show that one can stop the eight months of continual rainfall, and that one could prodigiously increase the heat in all this country, if one destroyed the forests that cover it, if one constrained the waters and directed the rivers, and if the cultivation of the land, which presumes the movement and great number of animals and of men, drove away the cold and superfluous humidity that the infinitely too-great quantity of vegetation attracts, maintains, and spreads widely?

Like all movement, all action produced by heat, and with all beings endowed with progressive movement being themselves nothing so much as little furnaces of heat, it is the proportion of the number of men and animals to that of plants that determines (all other things being equal) the local temperature of each terrain in particular. The former are a source of heat, and the latter produce only cold humidity. The habitual use that man has made of fire adds much to this artificial temperature in all those places that he inhabits in numbers. In Paris, in times of great cold, the thermometers in Faubourg Saint-Honoré show two to three degrees of cold more than at Faubourg Saint-Marceau, because the wind from the north is tempered on passing above the chimneys of this great city. A single forest more or less in a country suffices to change the temperature. While trees are around, they attract the cold, they diminish by their shade the heat of the Sun, they produce damp vapors, which form clouds and fall back as rain, all the colder as it has descended from height. And, if these forests are abandoned to Nature alone, these same trees, fallen from old age, decay coldly upon the earth, while in the hands of man, they serve to feed the element of fire, and become secondary causes of all particular heat. In the lands of the prairies, before the harvest of grasses, one always has abundant dew and very frequently little rains, which cease as soon as the grasses are harvested. These little rains would therefore become more abundant and would not cease, if our prairies, like the savannahs of America, were always covered with the same quantity of grasses which, far from diminishing, can only augment, by the fertilization from all those that dry and decay upon the earth.

I could easily give many other examples,[4] which all concur to show that man can modify the influences on the climate in which he lives, and can secure, so to speak, the temperature to the level at which it suits him. And here is something that is singular: it is more difficult for him to cool the Earth than to heat it. Master of the element of fire, that he can augment and propagate at his will, he cannot do the same with the element of cold, which he can neither grasp nor communicate. The principle of cold is not even a real substance, but a simple lack or rather a diminution of heat, a diminution that must be very large in the high regions of the air, and it is large enough at a league of distance from the Earth to convert aqueous vapors into hail and snow.

Because the emanations of the globe's own heat follow the same law that do all the other physical quantities or qualities that leave from a common center, and their intensity decreases in an inverse sense to the square of the distance, it seems certain that it is four times as cold at two leagues than at one league of height in our atmosphere, in taking every point at the Earth's surface as center. On the other hand, the interior heat of the globe is constant in all the seasons at ten degrees above freezing. In this way all greater cold, or rather all heat less than ten degrees, can come to the Earth only by the fall of matter chilled in the higher regions of the air, where the effects of this internal heat diminish as one goes higher. Now the power of man does not extend so far. He cannot make the cold descend as he can make heat rise; he has no other means to protect himself from the excessive heat of the Sun than by creating shade. But it is much easier to fell the forests in Guyana to warm the humid earth than to plant them in Arabia to refresh the arid sands. However, a single forest in the middle of these burning deserts is sufficient to make them more temperate, to bring there the waters of the sky, to bring to the Earth all the principles of its fertility, and consequently to make man enjoy there all the mildness of a temperate climate.

It is upon the difference in temperature that depends the greater or lesser energy of Nature: the growth, the development, and the very production of all the organized beings are only particular effects of this general cause. Thus man, in modifying that, can at the same time destroy that which harms him and bring forth all that is suitable for him. Happy are the countries where all the elements of temperature are in balance and advantageously enough combined to work only to good effect! But are there any that had this privilege from their origin? Any, where the power of man has not assisted that of Nature, either by bringing in or turning away water, or by destroying useless herbs and plants that are harmful or superfluous, or by favoring useful animals and multiplying them? Of three hundred species of quadruped animals and fifteen hundred species of birds that populate the surface of the Earth, man has chosen nineteen or twenty. These twenty species themselves figure more largely in Nature and make more good on the Earth than all the other species together. They figure more importantly because they are directed by man, who has prodigiously multiplied them. They operate in concert with him all the good that one can expect from a wise administration of force and of power for the cultivation of the Earth, for the transport and commerce of its productions, for the increase of materials, in a word, for all the needs and even for the pleasures of the sole master who can pay for their services by his care.

And in this small number of animal species that man has chosen, that of the chicken and the pig, which are the most fecund, are also the most generally widespread, as if the aptitude for the greatest multiplication was accompanied by this vigor of temperament that braves all hardships. One has found the chicken and the pig in all the least frequented parts of the Earth, on Tahiti and on all the other islands that have been longest unknown and that are the most distant from the continents; it seems that these species have followed that of man in all his migrations. On the isolated continent of South America where none of our animals could penetrate, the peccary and the wild chicken have been found which, although smaller and a little different from the pig and the chicken of our continent, must nevertheless be regarded as very closely related species that one could similarly reduce to domesticity; but savage humans having no idea of society, they did not even seek that of animals. In all the lands of South America, the savages have no domestic animals at all. They indiscriminately kill both good species and bad; they do not choose any to breed and to multiply, while a single species like the *curassow*, which they have at hand, would provide them, without effort and with only a little care, more subsistence than they can get for themselves by their arduous hunting.

Thus the first trait of man as he starts to become civilized is the empire that he learns to take over animals, and the first trait of his intelligence then becomes the greater character of his power over Nature. Because, it is only after having subjugated them that he has, through their help, changed the face of the Earth, converted the deserts into plowed land and the heaths into cornfields. In multiplying the species of useful animals, man has increased the amount of movement and of life on the Earth, he has at the same time ennobled the entire series of organisms and ennobled himself in transforming plant into animal and both into his own substance, which then spreads out by many multiplications. Everywhere that he produces in abundance, there always follows a great population; millions of people exist in the same space that was once occupied by two or three hundred savages, and thousands of animals where there were barely a few individuals. Through him and for him the precious germs are solely the developed ones, the most noble productions the cultivated ones alone. On the immense tree of fecundity, the fruit-bearing branches are the sole sustaining ones, all being perfected.

The grain with which man makes his bread is not a gift of Nature but the great, useful fruit of his researches and his intelligence in the first of the arts. Nowhere on Earth has wild wheat been found, and it is evidently a plant perfected by his efforts. He must thus have recognized and chosen, among thousands and thousands of others, this precious grass; he must have sown it, gathered it many times to note its multiplication, always proportional to the cultivation and to the fertility of the Earth. And this property, so to speak unique, that wheat has, to resist in its early stages the cold of our winters, although subject, like all the annual plants, to perish after giving its grain, the marvelous quality of this grain, which suits all people, all animals, almost in all climes, and which besides keeps a long time without alteration, without losing its power to reproduce, all this shows us that it is the happiest discovery that man has ever made, and however ancient one may suppose it to be, it was nevertheless preceded by the art of agriculture founded on science and perfected by observation.

If one wishes to have more modern, and even more recent examples of the power of man upon the nature of plants, one has only to compare our vegetables, our plants, and our fruits with the same species as they used to be a hundred and fifty years ago. This comparison can be made immediately and very precisely on running your eyes over the great collection of colored drawings, commenced at the time of *Gaston d'Orleans* and which still continues today in the Jardin du Roi. One will see there, perhaps with surprise, that the most beautiful flowers of that time, buttercups, carnations, tulips, bears'-ears and so on, would be rejected today, I do not say by our florists, but by the village gardeners. These flowers, although already then cultivated, were not yet far from their state of nature. A simple row of petals, long pistils, and with hard or false colors, without velvety texture, without variety, without nuances, with all the rustic characters of wild nature. Among the kitchen plants, a single species of chicory and two types of lettuce, both quite bad, while today we can count on more than fifty types of lettuce and chicory, all very good to the taste. We can similarly provide the very modern date of our best fruit with pips and with kernels, all different from those of the ancients, which they resemble only in name. Ordinarily things stay the same and names change over time; here it is the contrary, the names have remained and the objects have changed. Our peaches, our apricots, our pears are new productions for which one has conserved the odd names of earlier productions. To have no doubt of this, one need only compare our flowers and our fruits with the descriptions or rather the accounts that the Greek and Latin authors have left us; all their flowers were simple and all their fruit trees were only wild ones, badly enough chosen of each type, of which the little fruits, sharp or dry, had neither the flavor nor the beauty of ours.

It is not that there were no good or new species that did not originally stem from wild stock. But how many times were needed for man to tempt Nature to obtain from it the excellent species? How many thousands of germs was he obliged to entrust to the earth so that he would finally produce them? It is only in sowing, in raising, in cultivating and putting to fruit an almost infinite number of plants of the same species that he was able to recognize a few individuals bearing fruit that were softer and sweeter and better than the others. And this first discovery, which already implies much effort, would have yet remained sterile forever if there were not a second person, who must have had as much ingenuity as the first needed patience. This was to have found the means to multiply by grafting those precious individuals, which unfortunately could not produce a line as noble as them nor propagate by themselves their excellent properties. And this alone proves that in effect there are only purely individual qualities and not specific properties; because, the seeds or pips of these excellent fruits do not produce like others, except simple wild ones and consequently they do not form species that would be essentially different. But, by means of grafting, man could so to speak create secondary species that he could propagate and multiply at his will. The bud or the little branch that he joins to the wild form encloses this individual quality, which cannot be transmitted by seed, and which only needs to develop to produce the same fruit as the individual from which they were separated to unite them with the wild form, which does not communicate to them any of its bad qualities, because it did not contribute to their formation. It is not a mother, but a simple wet nurse, which only serves for their development through nourishment.

In the animals, most of the qualities that seem individual cannot be transmitted or be propagated by the same route as the specific properties. It was therefore easier for man to influence the nature of animals than that of plants. The races in each species of animals are only constant varieties that are perpetuated by generation, while in plant species there are no races at all, or any varieties constant enough to be perpetuated by reproduction. In the species of the chicken and pigeon alone, a great number of new races have been created very recently, which are all capable of propagating themselves. Every day, among other species, one elevates or ennobles the races in crossing them; from time to time foreign or wild species are acclimatized or cultivated. All these recent and modern examples prove that man has only lately known the extent of his power, and even that he does not yet know enough; it depends entirely upon the exercise of his intelligence. And so, the more he will observe, the more he will cultivate Nature, the more he will have the means to subjugate it and to make it easier to draw new riches from her heart, without diminishing the treasures of her inexhaustible fecundity.

And what could he not do upon himself, I wish to say upon his own species, if the will was always guided by intelligence? Who knows to what point man could perfect his nature, either moral or physical? Is there a single nation that can boast to have arrived at the best government possible, which would make all men not equally happy, but less unequally unhappy? In watching over their care, so to spare their sweat and their blood through peace, by the abundance of materials, by the ease of their life and the means of their propagation: here is the moral goal of all society that seeks to better itself. And for physics, for medicine, and the other arts of which the object is to preserve us, are they as advanced, as well known, as the arts of destruction, the children of war? It seems that man has always reflected less upon the good than he has strived for evil. All society is a mixture of one and the other; and like all the sentiments that affect the multitude, fear is the most powerful; the great talents in the art of making evil were the first to strike the mind of man, and then those that amused him occupied his heart, and it was only after long use of these two means to false honor and sterile pleasure that finally he recognized that his true glory is science, and that peace is his true happiness.

Justifying Notes to the Facts Reported in the *Epochs of Nature*

Notes on the First Discourse

1. *The inherent and interior heat of the Earth seems to increase as one descends.*

"It is not necessary to dig far before finding, at first, a constant and unvarying temperature, whatever is the temperature of the air at the surface of the ground. One knows that the liquid in the thermometer stays perceptibly the same, the year round at the same height, in the cellars of the observatory, that are, nevertheless, only eighty-four feet or fourteen toises in depth from the entrance. That is why one fixes at that point the average or moderate level of our climate. This heat is normally maintained at more or less the same level from a similar depth of fourteen or fifteen toises as far as sixty, eighty or one hundred toises and beyond, more or less, according to circumstances, as one experiences in mines. After this, it increases and sometimes becomes so great that the workmen would not know how to withstand it and to live there, if one did not provide them with some refreshments and with fresh air, either by *respiration wells*, or by waterfalls . . . M. de Gensanne experienced, in the mines of Giromagny, three leagues from Béfort, that a thermometer carried to fifty-two toises of vertical depth showed ten degrees, as in the cellars of the observatory; that at 106 toises of depth, it was at ten and a half degrees; at 158 toises, it rose to fifteen and a fifth degrees, and at 222 toises of depth it reached eighteen and a sixth degrees." *Dissertation sur la glace*, by M. de Mairan, Paris, 1749, intro.–12, 60ff.

"The more one descends into great depths in the interior of the Earth," said M. de Gensanne elsewhere, "the more one experiences a distinct warmth, which always increases the further one descends. This reaches such a point, that at 1,800 feet of depth below the floor of the Rhine, near Huningue in Alsace, I have found that the heat is already strong enough to make water perceptibly evaporate. One can see the details of my experiments on this subject in the last edition of the excellent *Traité de la glace* by my late illustrious friend M. Dortous de Mairan." *Histoire naturelle de Languedoc*, tome 1, 24.

"All the rich veins in all kinds of ore," said M. Eller, "are perpendicular fissures of the Earth, and one knows not how to determine the depth of these fissures. There are some in Germany where one descends to beyond 600 perches (lachters*): the deeper the miners descend, the hotter the air they always encounter." *Mémoire sur la génération des métaux*, Académie de Berlin, 1733.

* I am assured that the lachter is a measure about equal to the brasse of five feet in length; this gives a depth of 3,000 feet for these mines.

2. *The temperature of the water of the sea is approximately equal to that in the interior of the Earth at the same depth.*

Having submerged a thermometer in the sea at different places and at different times, it was found that the temperature at ten, twenty, thirty, and one hundred twenty fathoms was consistently from ten degrees to ten and three-quarters degrees. See *L'Histoire physique de la mer*, by Marsigli, page 16 . . . M. de Mairan has made a very judicious remark on this subject: "It is because the warmest waters, which are at the greatest depth, must, being lighter, be continually rising above those which are less so, which gives to this great liquid layer of the terrestrial globe a temperature that is more or less equal, conforming to the observations of Marsigli, except for the surface now exposed to the effects of the air, or where the water sometimes freezes before it has the time to descend by its weight and by its cooling." *Dissertation sur la glace*, 69.

3. *The light of the Sun does not penetrate more than to 600 feet of depth in the water of the sea.*

The late M. Bouguer, learned astronomer of the Royal Academy of Sciences, has observed that with sixteen pieces of the ordinary glass that windows are made with, applied one against the other, and making in all a thickness of nine-and-a-half lines, the light, passing through these sixteen pieces of glass, diminishes two hundred and forty-seven times, that is to say, that it was two hundred and forty-seven times weaker than before it had to traverse these sixteen pieces of glass. Then, he placed seventy-four pieces of this same glass at some distance one from another in a pipe, to diminish the light of the Sun to extinction: this star was at fifty degrees in height above the horizon when he made this experiment; and these seventy-four pieces still did not stop some semblance of its disc. Several people who were with him also saw a feeble glow, which they could only distinguish with difficulty, and which disappeared as soon as their eyes were no longer in complete darkness. But, when another three pieces of glass had been added to the seventy-four first ones, none of the assistants saw any light. Thus, in supposing eighty pieces of the same glass, a thickness of glass is obtained necessary for there no longer to be any transparency to allow even the most delicate images; and M. Bouguer found by an easy enough calculation that the light of the Sun is thus rendered 900 billion times more feeble; hence, all transparent matter which, by its great thickness, will cause the light of the Sun to be diminished 900 billionfold, will lose all of its transparency.

In applying this rule to the water of the sea, which of all waters is the most limpid, M. Bouguer found that, to lose all of its transparency, 256 feet of thickness are needed, given that, through another experiment, light from a torch had diminished in a scale from fourteen to five in traversing 115 inches of thickness of water contained in a canal nine feet and seven inches long, and that by an incontestable calculation, it must lose all transparency in 256 feet. Thus, according to M. Bouguer, detectable light cannot pass beyond 256 feet into the depths of water. *Essai d'Optique sur la gradation de la lumière*. Paris, 1729, 85, in-12.

However, it seems to me that this result of M. Bouguer is still far from reality. It would have been desirable if he had made his experiments with masses of glass of different thickness, and not with pieces of glass placed one above the other. I am persuaded that the light of the Sun would have penetrated a greater thickness than that of those eighty pieces which, all together, make up only forty-seven-and-a-half lines, that is to say, about four inches and, although these pieces that he used were of common glass, it is certain that a solid mass of four inches in thickness of the same glass would not have entirely intercepted the light of the Sun, to the extent that I am assured, by my own experience, that a thickness of six inches of clear glass lets light pass through it easily enough, as one will see in the following note. I thus think that one has to more than double the thicknesses given by M. Bouguer, and that the light of the Sun penetrates at least 600 feet through the water of the sea, because there is a second oversight in the experiments of this learned physicist, which was not to have allowed the light of the Sun to pass through his pipe filled with seawater, nine feet and seven inches long. He was content to use the light of a torch, and concluded a diminution in the relation of fourteen to five; and, I am persuaded that this diminution would not have been so great upon the light of the Sun, the more so in that the light of the torch could pass through it only obliquely; instead, that from the Sun passing overhead would have been more penetrating just from its incidence, independently of its purity and of its intensity. Thus, everything being considered, it seems to me that to approach as closely as possible to the truth, one must suppose that the light of the Sun penetrates in the heart of the sea up to 100 toises, or 600 feet, of depth, and that the heat as far as 150 feet. That is not to say that a few atoms of light or heat do not go further; but only that their effect would be absolutely imperceptible, and could not be recognized by any of our senses.

4. *The heat of the Sun might not penetrate further than 150 feet of depth into the water of the sea.*

I think myself assured of this truth by an analogy drawn from an experiment that seems to me decisive: with a lens of massive glass twenty-seven inches in diameter and six inches thick at its center, I saw, on covering its central part, that this lens burned, so to say, by its edges only up to four inches in thickness, and that all of its thicker part produced virtually no heat. Then, having covered all of this lens with the exception of an opening an inch across at its center, I saw that the light of the Sun was so diminished after having traversed this thickness of six inches of glass, that it did not produce any effect on a thermometer. I am thus well founded to presume that this same light, diminished by 150 feet of thickness of water, would not give a perceptible degree of heat.

The light that the Moon reflects to our eyes is certainly light reflected from the Sun; however, this light has no perceptible heat, and even when one concentrates it in the center of a burning mirror, which prodigiously increases the heat of the Sun, this light, reflected by the Moon, still has no perceptible heat at all; and that of the Sun will have no more heat once it has traversed a certain thickness of water, and will become as feeble as that of the Moon. I am thus persuaded that in letting the rays of the Sun pass through a large pipe filled with water, only fifty feet long, which is only a third of the thickness I have inferred, this enfeebled light would not produce any effect on a thermometer, even if we supposed the liquid in the thermometer to be at its freezing point. From this, I have thought to be able to conclude that although the light of the Sun penetrates up to 600 feet into the heart of the sea, its heat does not penetrate to a quarter of this depth.

5. *All the materials of the globe are of the nature of glass.* This general truth, which we can demonstrate by experiment, had been suspected by Leibnitz, a philosopher whose name will always give great honor to Germany.

"Of course, most people believe, and it was even suggested by the sacred writers, that there is a great store of fire hidden in the recesses of the earth. . . . . . appearances support this, because all scoriae created by fusion are a kind of glass. . . . . . But we can actually see that the surface of our globe was formed thus (and clearly it is not possible for us to penetrate further), for all earths and rocks revert to glass through the action of fire. . . . . For it suffices to assert that everything terrestrial is, through fire, turned to glass. The great bones of the earth, the naked cliffs and immortal sands are largely transformed into glass, as though condensed from bodies formerly melted by the great force of primordial fire that acted upon the tractable material of nature. . . . . and just as everything that does not escape as vapor will in the end be melted by the use of burning mirrors and acquire the nature of glass, so it is easily understood that glass is like the basis of the earth, and that its nature, for the most part, lies hidden behind the masks of other materials." Leibnitz, *Protogaea*, 1749, 4–5.

6. *All the terrestrial materials have glass as their base, and can be reduced to glass by means of fire.*

I admit that there are a few materials that the fires of our furnaces cannot reduce to glass but, by means of a good burning mirror, these same materials can be so reduced. Here is not at all the place to report the experiments made with the mirrors of my invention, of which the heat is great enough to volatilize or vitrify all the materials exposed to their focus. But it is true that to this day there are no mirrors powerful enough to reduce to glass certain types of vitrescible matter, such as rock crystal, silex, or flint. It is not that these materials cannot by their nature be reduced to glass like the others, but solely that they demand a fiercer fire.

7. *The bones and the tusks of these ancient elephants are at least as long and as thick as those of living elephants.*

One can be assured of this by the descriptions and dimensions given by M. Daubenton, volume XI of this *Histoire naturelle*, in the article on the elephant; but since that time I have been sent an entire tusk and some other fragments of fossil ivory, of which the dimensions greatly exceed the length and normal thickness of elephant tusks, I have even made searches at all the merchants in Paris that sell ivory, and no one has found a tusk comparable to that one, and only one has been found, within a very great number, equal to those that have come to us from Siberia, of which the circumference is nineteen inches at the base. The merchants term *raw ivory* that which has not been in the earth and that one takes from living elephants, or that is found in forests with recent skeletons of these animals; and they give the name *cooked ivory* to that taken from the ground, and of which the quality is more or less denatured by a greater or lesser length of burial, or by the more or less active quality of the earth in which it has been interred. Most of the tusks that have come to us from the north are still of a very solid ivory, with which one can make beautiful artifacts. The largest have been sent to us by M. de l'Isle, astronomer, of the Royal Academy of Sciences; he collected them on his journey through Siberia. In all the shops in Paris there was only one single tusk of crude ivory, which is nineteen inches in diameter; all the others were smaller. This large tusk was six feet and one inch in length, and it seems that those that are in the Cabinet du Roi, and that had been found in Siberia, were more than six-and-a half-feet long when they were entire, but as the extremities have been truncated, one can judge this only approximately.

If one compares the femurs, similarly found in the lands of the north, one will assure oneself that they are at least as long and considerably thicker than those of the elephants of today.

Moreover, we have, as I have said, exactly compared the bones and tusks that have been sent to us from Siberia with the bones and tusks of an elephant skeleton, and we have clearly seen that all these bones are the remains of these animals. The tusks that have come from Siberia not only have the shape, but also the true structure of elephant ivory, of which M. Daubenton has given a description in the following terms:

"When the tusk of an elephant is cut transversely, one sees at the center, or approximately at the center, a black point, which is called the *heart*; but if the tusk was cut in the area of its cavity, there is at the center only a round or oval hole. One sees curved lines, which extend in opposing senses, from the center to the exterior, and which cross, forming little lozenges; there is normally a narrow circular band around the exterior; the curved lines ramify away from the center, and the number of these lines becomes greater as they approach the exterior. Thus, the size of the lozenges is almost everywhere about the same; their sides, or at least their angles, have a more vivid color than the areas doubtless because their substance is more compact; the band of the exterior is sometimes composed of straight transverse fibers, which would end at the center if they were prolonged. It is the appearance of these lines and of these points that one considers to be the grain of ivory. One sees it in all ivory, but it is more or less visible in different tusks; and, among that ivory in which the grain is sufficiently apparent to give it the name of *granular ivory*, there is one that is called *large-grained ivory* to distinguish it from ivory of which the grain is fine" (see *l'Histoire naturelle*, tome XI, 123ff., and *Mémoires de l'Académie des Sciences* of 1762).

8. *The state of captivity alone would have reduced these elephants to a quarter or a third of their size.*

This is shown to us by the comparison that we have made of the entire skeleton of an elephant, which is in the Cabinet du Roi and which had lived seventeen years in the Ménagerie of Versailles, with the tusks of other elephants from their native country. This skeleton and these tusks, although considerable in size, are certainly smaller by half in volume than the skeletons and tusks of those that lived free, whether in Asia or in Africa, and at the same time they are at least two-thirds smaller than the bones of these same animals found in Siberia.

9. *One finds elephant tusks and bones not only in Siberia, in Russia, and in Canada, but also in Poland, in Germany, in France, in Italy.*

Independently of all the fragments that have been sent to us from Russia and Siberia, and which we keep in the Cabinet du Roi, there are several others in certain private collections in Paris. There are a great many in the museum of Petersburg, as one can see in the catalogue that was printed in the year 1742; there are similar ones in the museum of London, in that of Copenhagen, and in several other collections in England, in Germany, and in Italy; there have even been several works of art made with this ivory in the lands of the north. Thus, one cannot doubt the great quantity of these elephant remains in Siberia and in Russia.

M. Pallas, learned naturalist, found, on his travels in Siberia these last few years, a great quantity of elephant bones, and the entire skeleton of a rhinoceros, which was buried only a few feet down.

"Monstrous elephant bones have just been discovered in Swijatoki, seventeen versts from Petersburg. They have been pulled out of a terrain that has long been flooded. Thus one cannot further doubt the prodigious revolution that changed the climate, the products, and the animals in all the countries of the Earth. These natural medals prove that the lands that are today devastated by the rigors of cold once possessed all the advantages of the south," *Journal de politique et de littérature*, 5 January 1776, article "*Pétersbourg*."

The discovery of skeletons and of tusks of elephants in Canada is fairly recent, and I was informed of the first of these in a letter from the late M. Collinson, member of the Royal Society of London—here is the translation of this letter:

"M. George Croghan assured us that, in the course of his travels in 1765 and 1766 in the lands neighboring the Ohio River, about four miles southeast of this river, 640 miles distant from Fort Duquesne (which we now call Pittsburgh) he saw, in the neighborhood of a large salt marsh, where the wild animals assembled at certain times of the year, large bones and teeth, and having examined this place carefully he found, on a piece of higher ground at the side of the marsh, a prodigious number of bones of very large animals, which by the length and form of these bones and of these tusks must lead to the conclusion that these are the bones of elephants.

"But these large teeth that I send to you, Sir, found with these tusks, and others yet larger than these, seem to indicate and even demonstrate that they did not belong to elephants. How can one reconcile this paradox? Could not one suppose that there once existed a larger animal, which had the tusks of an elephant and the jaws of a hippopotamus? Because these great molar teeth are very different from those of the elephant. M. Croghan thinks, from the great quantity of these different kinds of teeth, that is to say, the tusks and the molar teeth that are observed at this place, that there were at least thirty of these animals. Yet, elephants were not at all known in America, and probably they could not have been brought there from Asia. The impossibility for them to live in these countries, because of the rigor of the winters, and where nevertheless there is found a great quantity of their bones, gives a further paradox, which your eminent sagacity must determine.

"M. Croghan sent to London, in the month of February 1767, the bones and the teeth that he had collected in 1765 and 1766.

"1. To My Lord Shelburne, two great tusks, of which one was intact and which was close to seven feet long (six feet and seven inches in France), the thickness was that of an ordinary tusk of an elephant, which would have this length.

"2. A jawbone with two molar teeth held within it, and in addition, several separate very large molar teeth.

"To Dr. Franklin. 1. Three elephant tusks, of which one is about six feet long, being broken in half, damaged or gnawed in the center, and in appearance like chalk; the others in a very good state, the end of one of the two being sharpened to a point and of a very beautiful ivory.

"2. A small tusk about three feet long, as thick as an arm, with the sockets that received the muscles and the tendons, which was of a shiny chestnut color, these seemingly as fresh as if one had pulled them out of the head of the animal.

"3. Four molars of which one of the largest was larger and had one more row of points than those that I sent you. You can be assured that all of those that had been sent to My Lord Shelburne and to M. Franklin were of the same form and had the same enamel as those that I place before your eyes.

"Doctor Franklin recently dined with an officer, who reported from that same place, neighboring the Ohio River, a whiter tusk, shinier, even, than all the others, and a jawbone yet bigger than all those that I have mentioned." Letter from M. Collinson to M. de Buffon, dated from Mill Hill, near London, 3 July 1767.

Extract from the travel journal of M. Croghan, made on the Ohio River, sent to M. Franklin, in the month of May 1765:

"We passed the great river of Miami, and in the evening, we arrived at the place where elephant bones have been found; this can be some 640 miles from Fort Pitt. In the morning, I went to see the large marshy area where wild animals assemble at certain times of year; we arrived at this place by a track made by wild oxen (*bisons*), some four miles distant from the southeast of the Ohio River. We saw with our eyes that there are in these places a great quantity of bones, some scattered, others buried to a depth of five or six feet below ground, which we saw by the thickness of the bank of earth that borders this type of trackway. There we found two tusks six feet long, which we took with us, with other bones and teeth; the following year, we returned to the same place, to take a greater number of other tusks and teeth.

"If M. de Buffon has doubts, and questions to ask on this, I ask you to tell M. Collinson to send them to me. I will pass his letter to M. Croghan, a very honest and enlightened man, who would be delighted to answer his questions." This small memoir was with the letter I have just cited, and to which I will add the extract of that which M. Collinson wrote to me previously, on the subject of these same bones found in America.

"It was around a mile and a half from the Ohio River, six monstrous skeletons were buried upright, having tusks five to six feet long, which were of the form and the substance of elephant tusks; they were thirty inches in circumference around the root; they thinned toward the point, but one cannot know quite how they were joined to the jaw, because they were broken into pieces. One femur of these same animals was found quite complete; it weighed a hundred pounds and was four-and-a-half feet long. These tusks and this thigh bone show that this animal was of prodigious size. These facts were confirmed by M. Greenwood who, having been in these places, saw the six skeletons in this salt marsh; moreover, he found, in the same place, large molar teeth that did not seem to belong to an elephant, but rather a hippopotamus; and he brought several of these teeth to London, among others two weighing together nine-and-a-quarter pounds. He said that the jawbone was close to three feet in length, and that it was too heavy to be carried by two people; he had measured the interval between the orbits of the two eyes, which was eighteen inches. One English lady, made prisoner by the savages and brought to this salt marsh to teach them how to make salt by making the water evaporate, said that she remembered, by a singular circumstance, having seen these enormous bones; she recounted that three Frenchmen, who were breaking nuts, were all three seated on just one of these large thigh bones."

A little while after having written me these letters, M. Collinson read, to the Royal Society of London, two small memoirs on the same subject, in which I found some facts additional to those just brought to you, adding a word of explanation where this is needed.

"The salt marsh where these elephant bones were found is only four miles distant from the banks of the Ohio River, but it is more than seven hundred miles distant from the nearest seacoast. There was a path opened up by wild oxen, large enough for two wagons, which led straight to the place of this large salt marsh where these animals assembled, as well as all kinds of deer and roebuck, in a certain season of the year, to lick the ground and drink the salt water . . . The elephant bones are found on some kind of levee or rather beneath the bank that surrounds and overlooks the marsh at five or six feet of height. One saw there a very great number of bones and teeth that belonged to several animals of a prodigious size; there were tusks more than seven feet in length, and which are of very beautiful ivory. One cannot thus doubt that they belonged to elephants; but that which was singular is that so far no one has found among these tusks any molar or grinding tooth of an elephant, but only a great number of large teeth, of which each has five or six rounded points, which could only have belonged to the same animal of a very great size, and these great square teeth do not resemble at all the molars of an elephant, which are flat and four or five times as long as wide, and so these great molar teeth do not resemble the teeth of any known animal."

What M. Collinson says here is very true; these great molar teeth differ completely from the grinding teeth of the elephant, and in comparing them to those of the hippopotamus, which these great teeth resemble by their square form, one will see that they differ from these also by their size, being two, three, and four times more voluminous than the largest teeth of ancient hippopotami similarly found in Siberia and in Canada, while these teeth are themselves three or four times larger than those of hippopotami living today. All the teeth that I have seen in four heads of these animals, which are in the Cabinet du Roi, have a grinding face shaped in the form of a clover, and those found in Canada and in Siberia have the same character and only differ in size; but these enormous teeth with great rounded points differ from those of the hippopotamus shaped in a clover form; they always have four, or sometimes five, rows, compared with which the largest hippopotamus teeth have only three, as one can see in comparing the figures of plates I, III, and IV, with those of plate V. It thus seems certain that these large teeth have never belonged to either an elephant or a hippopotamus; the difference in size, although enormous, would not stop me from regarding them as belonging to this latter species, if all of the characters of the form were similar, because we know, as I have just said, other square teeth, three or four times larger than those of hippopotami today, and which nevertheless have the same characters of form, and particularly the cavity in the shape of a clover on the grinding face are certainly the teeth of hippopotamus three times greater than those of which we have the heads; and it is these large teeth (plate V) that are truly the teeth of hippopotami, of which I have spoken, while I have said that they are found equally in these two continents, as well as elephant tusks; but that which is very remarkable is that not only have true elephant tusks, and true teeth of large hippopotami, been found in Siberia and in Canada, but one has similarly found there these much more enormous teeth with rounded points in four rows; I am thus able to say on this basis that this very great species of animal is lost.

M. le comte de Vergennes, minister and secretary of state, had the goodness to give me, in 1770, the largest of all these teeth, which is illustrated (plates I & II); it weighs eleven pounds and four ounces; this enormous molar tooth was found in Little Tartary in making a ditch. There were other bones that were not collected; among others, a femur bone of which there remained only a half that was entire, and the cavity of this half could contain fifteen Paris pints. M. l'Abbé Chappe, of the Academy of Sciences, brought us, from Siberia, another very similar tooth, but not as large, and that weighs only three pounds twelve-and-a-half ounces (plate III, fig.1 & 2). Finally, the greatest of those that M. Collinson sent me, and which is illustrated (plate IV), was found with similar others in America, near the Ohio River; and others, which have come to us from Canada, resemble them perfectly. One thus cannot doubt that, independently of the elephant and of the hippopotamus, of which one finds equally the remains in the two continents, there was also another animal common to the two continents, of a size superior to that of even the largest elephants. Because the square form of these enormous grinding teeth prove that they were present in number in the jaw of the animal, and if one supposes only six or even four on each side, one can judge the enormity of the head, which would have at least sixteen grinding teeth each weighing ten or eleven pounds. The elephant has only four, two on each side. They are flattened, they occupy all of the space of the jawbone, and these two strangely flattened molar teeth of the elephant surpass only by two inches the size of the largest square tooth of the unknown animal, which is twice as thick as those of the elephant. Thus, everything leads us to believe that this ancient species, which one has to consider as the first and the largest of all the terrestrial animals, subsisted in these first times only, and has not survived into our times, because an animal of a type larger than an elephant could not hide anywhere on Earth to the point of remaining unknown. And, besides, it is evident that by the very form of these teeth, by their enamel, and by the disposition of their roots, they have no relation to the teeth of cachalots or other cetaceans, and that they really belonged to a terrestrial animal of which the species was closer to that of the hippopotamus than to any other.

In a later part of the memoir that I have cited here, M. Collinson said that several members of the Royal Society knew as well as he did the elephant tusks that are found every year in Siberia, on the banks of the Ob and other rivers of that country. What system will one establish, he adds, with what degree of probability, to give reason for these deposits of these elephant bones in Siberia and in America? He finishes by giving the listing, the dimensions, and the weight of all the teeth found in the salt marsh of the Ohio River, of which the largest square tooth belongs to Captain Ourry, and weighs six-and-a-half pounds.

In the second small memoir of M. Collinson, read to the Royal Society of London, the tenth of December 1767, he said that, having noticed that one of the tusks found in the salt marsh had striae close to the large end, he had doubts as to whether these striae were or were not particular to the elephant species; to satisfy himself, he went to visit the shop of a merchant who traded in teeth of all types, and having carefully examined them, he found that there were as many tusks with striae at the larger end as without, and that consequently, there was no longer any difficulty in pronouncing that the tusks found in America were similar in all respects to the elephant tusks of Africa and Asia. But, as the large square teeth found in the same place have no relation to the elephant molar teeth, he thinks that they are the remains of some enormous animal that had the tusks of an elephant, with the molar teeth particular to its species, that then was of a size and form different from that of any known animal. See the *Philosophical Transactions of the year 1767*.

Since the year 1748, M. Fabri, who made long journeys in the north of Louisiana and in the south of Canada, informed me that he had seen the heads and the skeletons of a quadruped animal of enormous size, which the savages called the *father-of-oxen*, and that the femur bones of these animals were five and up to six feet in height. A little while afterward, and before the year 1767, some people in Paris had already received several of the large teeth of the unknown animal, some others of hippopotamus, and also the bones of elephants found in Canada. The number of these is too great for one to doubt that these animals formerly existed in the northern countries of America, as in those of Asia and of Europe.

But elephants existed also in all the temperate countries of our continent: I have already mentioned the tusks found in Languedoc near Simore, and those found at Comingues in Gascony; I must add the most beautiful and the largest of all, which was given lately for the Cabinet du Roi by M. le duc de la Rochefoucauld, whose zeal for the progress of the sciences is founded on the great knowledge that he acquired of all its disciplines. He found this fine fragment in visiting, with M. Desmarets of the Academy of Sciences, the countryside around Rome: this tusk is divided into five fragments that M. le duc de la Rochefoucald had collected. One of these fragments was purloined by the scoundrel who had charge of it, and there are only four left, which are about eight inches in diameter. In assembling them, they make a length of seven feet; and we know, through M. Desmarets, that the fifth fragment, which was lost, was of three feet, and thus one can be assured that the entire tusk must have been about ten feet in length. On examining the broken surfaces, we have found there all the characters of elephant ivory; only this ivory, altered by a long sojourn in the ground, has become light and friable like other fossil ivory.

M. Tozzetti, learned naturalist of Italy, reported that there were found, in the valleys of the Arno, bones of elephants and of other terrestrial animals in great quantity, scattered here and there in the beds of earth, and he said that one could conjecture that these animals were long ago indigenous animals of Europe, and above all in Tuscany. Extract of a letter of *Docteur Tozzetti*, *Journal étranger*, December 1755.

"There was found," said M. Coltellini, toward the end of November 1759, "on a piece of rural land belonging to the Marquis of Petrella, and situated in Fusigliano in the territory of Cortona, a fragment of elephant bone encrusted in large part with a stony matter . . . Nowadays no one can find similar fossil bones in our neighborhood anymore.

"In the collection of M. Galvotto Carazzi, there is another large fragment of petrified elephant bone found in the last few years in the neighborhood of Cortona, at a place called *la Selva* . . . Having compared these fragments of bone with a piece of elephant tusk shortly arrived from Asia, one found between them a perfect resemblance.

"M. l'Abbé Mearini brought me, last April, the entire jaw of an elephant that he had found in the district of Farneta, a village of this diocese. This jaw is petrified in large part and, above the two sides where the rocky encrustation rises to the height of an inch, has all the hardness of stone.

"Finally I owe to M. Muzio Angelieri Allicozzi, gentleman of this town, an almost entire femur of an elephant, that he discovered himself, on one of the pieces of land that he owns called *la Rota*, in the territory of Cortona. This bone, which is as long as a Florence fathom, is also petrified, especially in the highest extremity that one calls the head." Letter of M. Louis Coltellini, of Cortona, *Journal étranger*, July 1761.

10. *These great petrified spirals, of which some are several feet in diameter.*

The knowledge of all the petrifactions, of which one no longer finds living analogues, would suppose a long study and reflective comparison of all the species of petrifactions that have been found to date in the heart of the Earth; and this science is not yet well advanced. Nevertheless, we are assured that there are several of these types, such as the horns of Ammon, the orthoceratites, the lenticular stones or numismales, the belemnites, the Judaic stones, the anthropomorphites, and so on, that one cannot assign to any species now living. We have seen petrified horns of Ammon, two or three feet in diameter, and we have been assured by trustworthy witnesses that one larger than a millstone was found in Champagne, since it was eight feet in diameter and one foot thick. This was even offered to me, but the enormous weight of this mass, of about eight thousand pounds, and the great distance from Paris, prevented me from accepting this offer. We no longer know the species of animals to which these remains, that we have named, belong. But, these examples, and several others that I could cite, suffice to prove that there once existed in the sea several species of shell and of crustacean that no longer live. It is the same with some scaly fish. Most of those that have been found in slates and in some shales do not sufficiently resemble the fish known to us that we can say they are of this or that species; those that are in the Cabinet du Roi, perfectly conserved in masses of stone, similarly cannot be assigned to known species. It seems thus that in all the groups, the sea once nourished animals of which the species no longer exist.

But, as we have said, we have up to the present only a sole example of a lost species among the terrestrial animals, and it seems that this was the largest of all, without excepting even the elephant. And since these examples of lost species among terrestrial animals are much rarer than among marine animals, does this not seem to show that the formation of the former was subsequent to that of the latter?

Notes on the First Epoch

1. *On the matter of which the core of comets is composed.*

I have said in the article on the formation of the planets (*Histoire naturelle, générale et particulière, tome 1, article 1, "De la formation des planètes,"* 1774, 137), *that the comets are composed of very solid and very dense matter*. This cannot be taken as a positive assertion in general, because there must be great differences between the density of this or that comet, as there are between the densities of different planets. But, we will only be able to determine the relative difference in density between each of these comets when we will know their periods of revolution as perfectly as the periods of the planets are known. A comet of which the density would be only like the density of the planet Mercury, double that of the Earth, and which would have at its perihelion as much velocity as the comet of 1680, would perhaps be sufficient to draw out of the Sun all the quantity of material that composes the planets, because in this case the matter of the comet being eight times more dense than the solar matter, it would communicate eight times as much movement, and draw out one eight-hundredth part of the mass of the Sun, as easily as a body the density of which would be equal to that of solar matter could draw out one hundredth part.

2. *The Earth is raised beneath the equator and lowered beneath the poles, in the exact and precise proportion demanded by the laws of weight, combined with those of centrifugal force.*

I have supposed in my treatise on the formation of the planets, volume I, page 128, that the difference in the diameters of the Earth was in the proportion of 174 to 175, after the determinations made by our mathematicians who had been sent to Lapland and to Peru. But, because they had supposed a regular curve of the Earth, I warned, page 165, that this supposition was hypothetical, and consequently I did not at all stop at this determination. I thus think that one must prefer the relation of 229 to 230, which was determined by Newton, after his theory and his experiments with a pendulum, which seem to me to be much surer than the measurements. It is for this reason that in the *Memoirs*, in the hypothetical part, I have always supposed that the relation between the two diameters of the terrestrial spheroid were of 229 to 230. M. Doctor Irving, who accompanied M. Phipps on his journey to the north in 1773, made very exact experiments on the acceleration of a pendulum at 79 degrees 50 minutes, and he found that this acceleration was from 72 to 73 seconds in 24 hours, from which he concluded that the diameter at the equator to that of the Earth's axis is as 212 to 211. This savant added, with reason, that his result approached that of Newton, much more than that of M. de Maupertuis, who gave a relation of 178 to 179, and more also than that of M. Bradley, who, following the observations of M. Campbell, gives a relation of 200 to 201 between the two diameters of the Earth.

3. *The sea on the coasts neighboring the town of Caen in Normandy has constructed, and still constructs by its flow and ebb, a type of mud made up of thin layers, which forms daily from the sediment in the water.*

Each rising tide brings and spreads across all the bank an impalpable silt, which adds a new fine layer to the older ones, from which results over a succession of time a soft mudstone that is laminated.

Notes on the Second Epoch

1. *The rock of the globe and the high mountains in their interior as far as their summit are only made of vitrescible matter.*

I have said (volume 1, page 70, of *Histoire naturelle*) "that the terrestrial globe could be empty in its interior, or full of a material more dense than all that we know, without us being able to demonstrate this, and we are barely able to make some reasonable conjectures about this." But when I wrote the treatise *Histoire naturelle* in 1744, I was not instructed in all the facts by which one could recognize that the density of the Earth, taken in general, is intermediate between the density of iron, of marble, of sand, of rock, and of glass, such as I determined it in my first *Memoir* (hypothetical part, supplement, tome II). I had not, though, made all the experiments that led me to this result. I also lacked the many observations that I have collected over this long span of time. These experiments were all made with the same aim, and these new observations, for the most part, have extended my first ideas and have led to the birth of other accessory ones, even more developed, with the result that these *reasonable conjectures*, that I had then suspected could be made, seem to me to have become very plausible inductions, from which it results that the globe of the Earth is mainly composed, from the surface to its center, of a vitreous matter a little denser than pure glass; the Moon, of matter as dense as limestone rock; Mars, of matter about as dense as marble; Venus, of a matter a little more dense than emery; Mercury, of matter a little more dense than tin; Jupiter, of matter less dense than chalk; and Saturn, of matter almost as light as pumice; and finally, that the satellites of these two large planets are composed of matter yet lighter than their principal planet.

It is certain that the center of gravity of the globe, or rather of the terrestrial spheroid, coincides with the center of its bulk, and that the axis upon which it turns passes through these same centers, that is to say, through the middle of the spheroid, and by that consequence, it is of the same density in all its corresponding parts. If it were otherwise, and if the center of its size did not coincide with the center of gravity, the axis of rotation would find itself then more on one side than the other, and in the different hemispheres of the Earth, the duration of revolution would seem unequal. And this revolution is perfectly the same for all the climes; thus, all of the corresponding parts of the globe are of the same relative density.

And when it is demonstrated—by its expansion at the equator and by its own heat, still present today—that at its origin the terrestrial globe was composed of matter liquefied by fire, which it assembled by the force of its mutual attraction, the union of this material in fusion could only form a full sphere, from the center to its circumference, which only differs from a perfect globe by its expansion beneath the equator and its depression under the poles, produced by centrifugal force from the first moments when this still-liquid mass began to rotate.

We have shown that the result of all matter that undergoes the violent action of fire is the state of vitrification; and as everything is reduced to more or less dense glass, the interior of the globe must in effect be of vitreous matter, of the same nature as vitreous rock, which makes everywhere the foundation of its surface below the shales, the vitrescible sands, the calcareous rocks, and all the other materials that have been stirred, worked, and transported by the waters.

Hence the interior of the globe is a mass of vitrescible material, perhaps specifically a little more dense than vitreous rock, in the fissures of which we search for metals. But, this is of the same nature, and differs only in that it is more massive and more solid. There are only spaces and caverns in the external layers; the interior must be solid, because caverns could form only at the surface at the time of consolidation and first cooling. The vertical fissures that are found in the mountains were formed almost at the same time, that is to say, while the matter was contracting through cooling. All the cavities could be formed only at the surface, as one sees in a mass of glass or of mineral that is melted; the elevations and depressions form at the surface, while the interior of the mass remains solid and full.

Independently of the general cause of the formation of the fissures and caverns on the surface of the Earth, centrifugal force was another cause that, combining with that of cooling, produced at first the largest caverns and the greatest inequalities in those climes where it acted most powerfully. It is for this reason that the highest mountains and the greatest depths are found neighboring the tropics and the equator; it is for this same reason that there are more upheavals in these southern countries than anywhere else. We cannot determine the depth to which the layers of the Earth have been blistered by fire and raised into caverns; but it is certain that this depth must be much greater at the equator than in other climes, because the globe, before its consolidation, was elevated there by six-and-a-quarter leagues more than beneath the poles. This type of crust or cover always diminishes in thickness away from the equator and dies away to nothing beneath the poles. The matter that composes this crust is the only one that had been displaced at the time of liquefaction, driven back by the action of centrifugal force; the rest of the material that makes up the interior of the globe stayed fixed in place, and did not undergo either change, or uplift, or transport. Spaces and caverns could thus form only in this external crust; they are larger and more numerous where this crust is thicker, that is to say, nearer the equator. The greatest founderings also took place and still take place in the southern parts, where are similarly found the greatest irregularities of the surface of the globe, and for the same reason, the greatest number of caverns, of fissures, and of metal ores, which then filled the fissures in the times of their fusion or of their sublimation.

Gold and silver, which make up only a quantity that is, so to speak, infinitely small in comparison with those of other materials of the globe, were sublimated in vapors, and were separated from the common vitrescible material by the action of heat, in the same way that one can see leaving, from a pool of gold or silver exposed to the furnace of a burning mirror, particles that are separated by sublimation, and that gild or silver the bodies to which one exposes this metallic vapor. Thus, one cannot believe that these metals, susceptible to sublimation even by moderate heat, could have entered in large part into the composition of the globe, nor that they were placed at great depths within its interior. It is the same with all the other metals and minerals, which are even more susceptible to sublimation by the action of heat. And, with regard to vitrescible sand and clays, which are only the detritus of vitreous scoriae, of which the surface of the Earth was covered immediately after the first cooling, it is certain that they could not lodge in its interior, and that they penetrate at most as deep as the metallic veins, in the fissures and in the other cavities of this ancient surface of the Earth, now covered by all the materials that the waters have deposited.

We are thus well founded to conclude that the globe of the Earth is in its interior only a solid mass of vitrescible matter, without spaces, without cavities, and there are found there only those layers that support those of the surface; that beneath the equator and in southern climes, these cavities were and still are larger than those in temperate or northern climes, because there are two causes that have produced them beneath the equator: to wit, centrifugal force and cooling; while beneath the poles there was only the sole cause of cooling, with the result that in southern parts, the founderings were much more considerable, the irregularities greater, the perpendicular fissures more frequent, and the ores of precious metal more abundant.

2. *The fissures and cavities of the high areas of the terrestrial Earth were encrusted, and sometimes filled, by metallic substances that we find there today.*

"The metallic veins," says M. Eller, "are found only in the high places, in a long line of mountains; this chain of mountains always implies for its support a base of hard rock. Inasmuch as this rock preserves its continuity, there is hardly anywhere where one can discover a few metal veins; but when one encounters some crevasses or fissures, one can hope to discover them. The mineralogist physicists have said that in Germany, the most favorable situation is where the mountain chains become higher little by little, are directed to the southeast and, having attained their highest elevation, descend insensibly toward the north west . . .

"It is generally a *native rock*, of which the extent is sometimes almost without boundaries, but which is split and opened up in various places, that contains metals that are sometimes pure, but nearly always mineralized. These fissures are generally coated with white and shiny earth, which the miners call *quartz*, and which they call *spath* when this earth is heavier, but flimsy and leaf-like, somewhat like talc; it is enveloped externally against the rock with a type of silt, which seems to provide nourishment for the quartz and spath-like earths. These two envelopes are like a sheath on the casing of the vein; the nearer they are to vertical, the more one is hopeful; and every time the miners see that the vein is perpendicular they say that it will become noble.

"The metals are formed in all these fissures and caverns by an evaporation that is continual and fairly violent; the vapors in the ores demonstrate this evaporation still continues. The fissures that do not exhale at all are generally sterile. The surest sign that the exhaling vapors carry the atoms or molecules of the mineral, and that they apply them everywhere on the walls of the crevasses of the rock, is this successive encrustation that one sees on all of the circumference of these fissures or cracks in the rocks, until its capacity is completely filled and the vein is solidly formed. This is further confirmed by the tools that are forgotten in the fissures, and later recovered, several years later, covered and encrusted in the ore.

"The fissures in the rock that furnish an abundant metallic vein are always inclined or approach being perpendicular to the ground. The further the miners descend, the warmer the temperature of the air they encounter, and sometimes exhalations are so abundant and so injurious to the respiration that they are forced to retreat as quickly as possible toward the shafts or toward the gallery, to avoid the suffocation that the sulfurous and arsenical parts would immediately cause. Sulfur and arsenic are generally found in all the ores of the four imperfect metals and of the half-metals, and it is by these that they are mineralized.

"There it is only gold, and sometimes silver and copper, that are found native in small quantities. But, generally, copper, iron, lead, and tin, while they are drawn from the lodes, are mineralized with sulfur and arsenic. One knows by experience that the metals lose their metallic form at a certain degree of heat relative to each species of metal. This destruction of the metallic form, which is undergone by the four imperfect metals, teaches us that the bases of the metals are of terrestrial matter; and as the metallic limes vitrify at a certain degree of heat, as do the calcareous earths, gypsums, and so on, we cannot doubt that the metallic earth consists of a number of vitrifiable earths." Extract of *Mémoire de M. Eller, sur l'origine et la génération des métaux*, in the *Recueil de l'Académie de Berlin*, 1753.

3. *M. Lehman, the celebrated chemist, is the only one to have suspected a double origin for the metallic ores: he has judiciously distinguished the mountains with veins from the layered mountains.*

"Gold and silver," he says, "are found in masses only in the 'mountains with veins.' Iron is hardly found except in the layered mountains. All the fragments or the little parcels of gold and of silver that one finds in the layered mountains are only scattered there, and are detached from the veins that are in the higher mountains neighboring these beds.

"Gold is never mineralized; it is always found native or virgin, that is to say, altogether formed in its matrix, although it is often there so disseminated in particles so slender, that one can search vainly to recognize it, even with the best microscopes. One finds no gold at all in the layered mountains; it is rare enough that silver is found there. These two metals belong by preference in the mountains with veins; one has nevertheless sometimes found, in slate, silver as little flakes or in the form of hairs; it is less rare to find native copper in slate, and commonly this native copper is also in the form of threads or of hairs.

"Iron ores recur a few years after having been dug; they are not found at all in the mountains with veins, but in the layered mountains; one has not at all yet found native iron in the layered mountains or, at least, this is a very rare thing.

"As to native tin, none exists by nature without recourse to fire; and this is also very doubtful for lead, although it is said that there are grains of lead in Massel in Silesia, as native lead.

"One finds native, flowing mercury in beds of oily clayey earth, or in slates.

"The silver ores found in slates are not nearly as rich as those found in the mountains with veins. This metal is hardly ever found except as slender particles, or as threads or growths, in these beds of slate or shale, but never as large ores; and furthermore these beds of slate need to be neighboring mountains with veins. None of the silver ores that are found in these beds are in a solid and compact form; all the other ores that contain silver in abundance are found in mountains with veins. Copper is found abundantly in beds of slate, and sometimes also in coal in the ground.

"Tin is the metal that most rarely occurs scattered in beds; lead is found there more commonly; one finds it in the form of galena, attached to slates, but it is found only very rarely with coal in the ground.

"Iron is almost universally distributed, and is found in beds, in a large number of different forms.

"Cinnabar, cobalt, bismuth, and calamine are also found commonly enough in beds." *Lehman*, tome III, 381ff.

"Coal, jet, amber, aluminous earth, were produced by plants, and above all by resinous trees that were buried in the heart of the Earth, and that suffered more or less decomposition; because one very often finds wood above the coal seams that has not been at all decomposed and becomes more so as it is buried more deeply in the ground. The slate, which forms the roof or the cover of the coal, is often full of the imprints of plants, which are normally found in forests, such as bracken, maidenhair, and so on, and what is remarkable here is that these plants, of which the imprints are found, are all foreign, and the trees also seem to be foreign trees. The amber, which may be regarded as a plant resin, often encloses insects that, looked at closely, do not at all belong to the climate where one finds them today. Finally, the aluminous earth is often flake-like, and resembles wood, sometimes more, sometimes less decomposed.

"Sulfur, alum and sal ammoniac are found in beds formed by volcanoes.

"Petrol and naphtha indicate a fire presently lit beneath the ground, which, so to speak, puts coal into distillation. There are examples of these subterranean conflagrations, which act only silently in coal mines, in England and in Germany. These burn for a long time without explosion, and it is in the surroundings of these underground fires that one finds hot thermal waters.

"The mountains that contain veins never contain coal, nor bituminous and combustible substances; these substances are never found except in the layered mountains." *Notes sur Lehman*, by M. le baron d'Olbac, tome III, 435.

4. *There are, in the countries of our north, entire mountains of iron, that is to say, of a vitrescible and ferruginous rock, etc.*

I will cite for example the iron ore near Taberg in Smaland, part of the isle of Gotland in Sweden. It is one of the most remarkable of these ores, or rather of these mountains of iron, which all have the property of submitting to the attraction of a magnet, which proves that they were formed by fire. This mountain is within a soil of extremely fine sand; its height is more than 400 feet, and its circumference is a league; it is entirely composed of a very rich ferruginous matter, and one even finds native iron there—another proof that it has experienced the action of fierce fire. On being broken, this ore shows small brilliant particles at the surface, which in places crisscross and in others are arranged as scales. The small crags that neighbor it most closely are of a pure rock (*saxo puro*). This one has been worked for around two hundred years; it is exploited for cannon powder, and the mountain seems very little diminished, except in the wells that are at the foot of the valley.

It seems that this ore does not have any regular beds; neither is the iron everywhere of the same quality. The whole of the mountain has many fissures, in places perpendicular and in others horizontal; they are all filled with sand that contains no iron. This sand is as pure and of the same type as that which borders the sea. Animal bones and stag horns are sometimes found in the sand, which shows that it was brought there by water, and it was only after the mountain of iron was formed by fire that sand filled the crevasses and the vertical and horizontal fissures.

The masses of ore that are extracted fall immediately to the foot of the mountain, by comparison with other ores where one has to draw the mineral from the entrails of the Earth. One must crush and roast this ore before putting it into the furnace, where it is melted together with limestone rock and charcoal.

This hill of iron is situated in a high mountainous region, some eighty leagues from the sea; it seems that it was formerly entirely covered by sand. Extract of an article in the periodical work that has as title: *Nordische beytrage, etc. Contribution du nord pour le progrès de la physique, des sciences et des arts*. A. Altone, chez David Ifers, 1756.

5. *There are mountains of lodestone in several countries, and particularly in those of our north.*

One has seen, for example cited in the previous note, that the mountain of iron of Taberg rises to more than 400 feet above the surface of the Earth. M. Gmelin, in his journey in Siberia, assures us that in the northern countries of Asia, nearly all the metal ores are found at the surface of the Earth, while in other countries, they are found deeply buried in its interior. If this fact is generally true, this would be a new proof that these metals were formed by primitive fire, and because of the globe of the Earth having a smaller thickness in its northern parts, they formed there closer to the surface than in the southern countries.

The same M. Gmelin has visited the large mountain of lodestone that is in Siberia, in the Baschkirs; this mountain is divided into eight parts, separated by valleys: the seventh of these parts produces the best lodestone. The summit of this part of the mountain is formed of a yellowish rock, which seems to have the nature of jasper. There are stones there that one would take from afar to be sandstone, which weigh two thousand, five hundred or three thousand pounds, but which have all of the character of lodestone. As they are often covered with moss, they do not attract iron and steel at a distance of more than one inch; those surfaces that are exposed to the air have the most magnetic quality and those that are buried in the earth have much less. Those parts that are most exposed to the attack of the air are softer and consequently are the least suitable to be worked: a large lump of lodestone at the size just stated is composed of quantities of smaller pieces of lodestone, which act in different directions. To work them well, it is necessary to separate them by sawing, so that all of the fragment that includes the quality of each particular lodestone conserves its integrity. One would seem to obtain, by this means, lodestones of great force. But, if one cuts the fragments randomly, there are many that are worthless, either because one is working a piece of rock that has no magnetic quality, or that encloses only a tiny amount, or because within one fragment there are two or three lodestones conjoined. In truth, these fragments do possess a magnetic quality, but as it is not directed toward the same point, it is not surprising that the effect of such a lodestone is subject to many variations.

The lodestone of this mountain, with the exception of that which is exposed to the air, is of a great hardness, stained black, and full of tuberosities that have small angular parts, as one commonly sees at the surface of a bloodstone, from which it differs only in color. But often, in place of these angular parts, one sees only a kind of ochre earth. In general, the lodestones with the small angular parts have less quality than the others. The place on the mountain where the lodestones are is almost entirely composed of good iron ore, which is extracted in small fragments between the lodestones. All of the highest section of the mountain encloses a similar ore; but the lower one descends, the less metal it contains. Lower down, below the lodestone ore, there are other ferruginous rocks, but which yield very little iron, if one wished to smelt them. The fragments that one extracts from it have the color of metal and are very heavy; they are unequal internally and have almost the semblance of scoria. These fragments on the exterior resemble lodestone closely enough; but those that are extracted from eight fathoms below the rock no longer possess any quality. Between these stones, there are other rock fragments that seem to be composed of very small particles of iron. The stone itself is heavy, but very soft; the particles inside resemble burnt material, and they have little or no magnetic quality. One also finds from time to time a brown iron mineral in beds an inch thick, but it yields little metal. Extract of *l'Histoire générale des voyages*, tome XVIII, 141ff.

There are several other lodestone mines in Siberia, in the Poïas Mountains. Ten leagues from the road that leads from Catherinbourg to Solikamskaia, is the mountain Galazinski; it is more than twenty toises high, and it is entirely of lodestone rock, brown in color, of hard and compact iron.

Twenty leagues from Solikamskaia, there is a cubic and greenish lodestone. The cubes have a brilliant shine: when one pulverizes them, they decompose into brilliant fire-covered flakes. Furthermore, lodestone is found only in chains of mountains oriented from south to north. Extract of *l'Histoire générale des voyages*, tome XIX, 472.

In the neighboring lands of the area of Lapland, at the limits of Bothnia, two leagues from Cokluanda, one can see an iron ore from which one can extract lodestones that are wholly good: "We admire with much pleasure," said the narrator, "the surprising effects of this stone, while it is still in its birthplace. It is necessary to use much force to extract stones as large as we wish to have; and the hammer used, which was the size of a thigh, remained so fixed on falling on the chisel that was in the rock, that the person who was striking had need of help to pull it away. I wished to experience this myself, and having taken hold of a large iron crowbar similar to those that are used to move the heaviest objects, and that I had trouble lifting, I approached the chisel, which attracted it with extreme force and held it with inconceivable firmness. I placed a compass in the middle of the hole where the ore was, and the needle turned continually at an unbelievable speed." (*Oeuvres de Regnard*, Paris, 1742, tome I, 185).

6. *The highest mountains are in the torrid zone, the lowest in the cold zones; and one cannot doubt that since the origin, the areas near the equator were the most irregular and the least solid of the globe.*

I have said (in *Histoire naturelle*, tome 1, article 1, p. 94) "that the mountains of the north are only hills by comparison with those of southern countries, and that the general movement of the waters produced these great mountains in a direction from the east to the west on the ancient continent, and from the north to the south on the new." When I composed, in 1744, this treatise on *l'Histoire naturelle*, I was not as educated as I am now, and the observations had not been made by which one recognized that the summits of the highest mountains are composed of granite and of vitrescible rocks, and that one does not find any shells on many of their summits. This proves that these mountains were not produced by the waters, but by primitive fire, and that they are as ancient as the time of the consolidation of the globe. All the peaks and cores of these mountains being composed of vitrescible matter, similar to the interior rock of the globe, they are equally the work of primitive fire, which first established these mountain masses and formed the great inequalities of the surface of the Earth. The water did its work only subsequently, after the fire, and could act to the height only where it found itself after the entire downfall of the waters of the atmosphere and the formation of the universal sea, which deposited, successively, the masses of shells that it nourished and the other materials that it washed in: that which formed the beds of clay and of calcareous matter that compose our hills and that envelop the vitrescible mountains up to a great height.

Furthermore, while I have said that the mountains of the north are only hills in comparison to the mountains of the south, this is only true in general; because there are in the north of Asia large areas of terrain that seem to be greatly elevated above the level of the sea. And, in Europe, the Pyrenees, the Alps, the Carpathians, the mountains of Norway, the Riphean and Rymnic mountains, are high mountains; and all of the southern part of Siberia, although composed of vast plains and modest mountains, seems to be yet higher than the summit of the Riphean mountains. But, these are perhaps the sole exceptions that one can make here: because, not only are the highest mountains found in climes nearer the equator than the poles, but it seems that it is in the southern climes that occur the greatest internal and external upheavals, as much by the effect of centrifugal force, at the first times of consolidation, as by the more frequent action of the subterranean fires, and the more violent movement of flux and reflux in subsequent times. Earthquakes are so frequent in southern India that the natives of this country do not give another name to the all-powerful Being other than that of the *shaker of the Earth*. The whole of the Indian archipelago seems only to be a sea of active or extinct volcanoes: one thus cannot doubt that the irregularities of the globe are greater near the equator than toward the poles. One can even be assured that this surface of the torrid zone has been entirely convulsed, from the eastern coast of Africa as far as the Philippines, and yet much farther into the southern sea. The whole of this coast seems to be only the remains, as debris, of a vast continent, of which all of the lowlands have been submerged. The action of all of the elements is united for the destruction of most of these equinoctial lands; because, independently of the tides, which are more violent there than on the rest of the globe, it seems that there are also more volcanoes, because these still subsist on most of these islands, of which a few, such as the islands of Mauritius and of Réunion, have been destroyed by fire and completely deserted since their discovery.

Notes on the Third Epoch

1. *The waters covered all of Europe as far as 1,500 toises above the level of the sea.*

We have said (in *Histoire naturelle,* tome 1, article 1, page 77) "that the entire surface of the Earth now inhabited was once below the waters of the sea; that these waters were higher than the summits of the highest mountains, since one finds on these mountains, as far as their summit, marine products and shells."

This demands an explanation, and even a few reservations. It is certain and recognized by thousands upon thousands of observations, that there are shells and other products of the sea on all of the surface of the Earth that is currently inhabited, and even on the mountains, to a very great height. I have put forward, upon the authority of Woodward, who was the first to collect these observations, that these shells can be found also to the summits of the highest mountains; especially as I was assured by my own and by other recent observations, that these occur in the Pyrenees and the Alps to 900, 100, 1,200, and 1,500 toises in height above the level of the sea, that they are similarly found in the mountains of Asia, and finally in the Cordillera in America, there has been recently found a layer at more than 2,000 toises above the level of the sea.*

* M. le Gentil, of the Academy of Sciences, communicated to me in writing, on 4 December 1771, the following fact: Don Antonio de Ulloa, he said, entrusted to me, in passing through Cadiz, to deliver from him to the Academy, two fossil shells that he extracted in the year 1761 from a mountain where there is quicksilver, in the district of Ouanca-Velica in Peru, of which the southern latitude is between thirteen and fourteen degrees. In the place where the shells were collected, the mercury level was seventeen inches and one-and-one-quarter lines, which corresponds to 2,222 and 1/3 toises in height above the level of the sea.

On the highest part of the mountain, which is among the highest in this canton, the mercury level was sixteen inches and six lines, which corresponds to 2337 and 2/3 toises.

In the town of Ouanca-Velica, the mercury level was eighteen inches and a half line, which corresponds to 1,949 toises.

Don Antonio de Ulloa told me he detached these shells from a hard and thick bed, of which he did not know the extent, and he was presently working on a memoir relating to these observations; these shells are a type of scallop.

One can thus not doubt that, in all the different parts of the world, up to the height of 1,500 or 2,000 toises above the level of today's sea, the world was covered by water, for a long enough time to produce shells there and to let them multiply; for their quantity is so considerable that their debris forms layers many leagues in length, of several toises of thickness across an undefined extent. In this way, they compose a fairly considerable part of the external layers on the surface of the globe; that is to say, all the calcareous matter which, as one says, is very common and very abundant in many countries. But above these highest parts, that is to say, above 1,500 to 2,000 toises of height, and often lower, one sees that the summits of many mountains are composed of quartz, of granite, and of other vitrescible materials produced by primitive fire, which contain in effect neither shells, nor madrepores, nor anything connected to calcareous matter. One can thus infer from this that the sea did not attain, or at least submerged for a very short time only, the highest parts and the most elevated peaks of the surface of the Earth.

As the observation of Don Ulloa, which we have just cited regarding shells found in the Cordillera, may still seem doubtful, or at least isolated in forming a single example, we must seek the support of his witness, that of Alphonse Barba, who said that in the middle of the most mountainous part of Peru, one finds shells of all sizes, some concave and other convex, and finely imprinted. Thus America, like the other parts of the world, was likewise covered by the waters of the sea; and if these first observers believed that one would find no shells at all on the mountains of the Cordillera, it is that the mountains, the highest on the Earth, are for the most part volcanoes that are still active, or extinct volcanoes which, by their eruptions, covered all of the adjacent lands with burnt matter. This not only buried, but also destroyed all the shells that could have been found there. It would thus not be astonishing that one would not encounter any marine products around these mountains, which are today or which once were burnt; because the terrain that surrounds them can be composed only of cinders, of scoria, of lava, and of other burnt or vitrified materials. Thus, there is no basis for the opinion of those who suggest that the sea did not cover the mountains, other than there are many of their summits where no shell or other marine product can be seen. But as one finds, in an infinite number of places and at up to 1,500 to 2,000 toises in height, shells and other products of the sea, it is evident that there were few mountain peaks or crests that were not overtopped by the waters, and that the places where one finds no shells show only that the animals that produced them did not live there, and that the movements of the sea brought none of their debris or their products there, as they brought them to all of the rest of the surface of the globe.

2. *The species of fish and of plants that live and vegetate in hot waters, up to 50 or 60 degrees on the thermometer.*

There are several examples of plants that grow in the hottest thermal waters, and M. Sonnerat has found fish in water of which the heat was so great, that he could not plunge his hand into it. Here is an extract of his relation on the subjection: "I found," he said, "two leagues from Calamba, on the Isle of Luçon near the village of Bally, a stream in which the water was hot, to the point at which the thermometer, in the scale of Réaumur, plunged into this stream a league from its source, still showed sixty-nine degrees. I imagined on seeing such a degree of heat that all of the production of Nature must be dead by the sides of this stream, and I was very surprised to see there very vigorous shrubs, their roots in the boiling water and their branches surrounded by the vapor; this was so considerable that the swallows that dared cross this stream at a height of seven or eight feet fell in there without further movement; one of these three shrubs was an *agnus castus* and the two others, *aspalatus*. During my stay in this village, I drank only water that I had cooled from this stream; its taste seemed to me earthy and ferruginous. Different baths had been constructed along this stream, of which the degrees of heat were proportional to the distance from the source. My surprise redoubled when I saw the first bath: fish swam in this water, in which I could not plunge my hand. I did everything that I could to obtain a few of these fish, but their agility and the clumsiness of the local people did not allow me to take even one. I examined them as they swam, but the vapor from the water did not allow me to distinguish them well enough to compare them with other types; however, I recognized this for the fish with brown scales: the length of the largest was four inches. I do not know how the fish came to be in these baths." M. Sonnerat supports his account with the testimony of M. Prevost, commissaire of the marine, who traveled with him across the interior of the island of Luçon. Here is how the testimony ends: "You are right, Monsieur, to give to M. Buffon the observations that you have assembled during this journey that we have made together. You wish that I confirm by writing that which so surprised us in the village of Bally, sited at the edge of the Laguna de Manille, at Los-Bagnos: I am annoyed that I do not have here a record of our observations made with the thermometer of M. de Réaumur; but I recall very well that the water of this little stream, which passes through this village to flow into the lake, caused the mercury to rise to sixty-six or sixty-seven degrees, at only a league from its source: the edges of this stream were lined with a grass that was always green. You surely have not forgotten this *agnus-castus*, which we have seen in flower, the roots of which were soaked in the water of this stream, and the trunk continually enveloped in mist that came out of it. The Franciscan father, parish priest of this village, assures me that he also saw fish in this same stream. As to me, I cannot vouch for this; but I have seen some in one of the baths, in which the heat made the mercury rise to forty-eight and fifty degrees. This is what you can testify to with confidence. *Signed* PREVOST." *Voyage à la Nouvelle Guinée*, by M. Sonnerat, correspondent of the Academy of Sciences and the Cabinet du Roi. Paris, 1776, 38ff.

I do not know whether fish have been found in our thermal waters, but it is certain that even in the warmest ones, the floor of the terrain is carpeted in plants. M. l'Abbé Mazeas has expressly said that, in the almost boiling water of the Solfatara of Viterbe, the floor of the basin is covered in the same plants that grow in the bottom of the lakes and swamps. *Mémoires des savans étrangers*, tome V, 325.

3. *It seems from the monuments that remain to us that there were giants among several types of animals.*

The large teeth with rounded points of which we have spoken indicate a gigantic species relative to other species, and even to that of the elephant; but this gigantic species no longer exists. Other large teeth, of which the grinding face is in the form of a *clover*, like those of the hippopotamus, and which nevertheless are four times larger than those of the hippopotami living today, show that there once were very gigantic individuals among the hippopotamus species. Enormous femurs, larger and much thicker than those of our elephants, show that the same is true for elephants: and we can cite several more examples that give support to our opinion concerning gigantic animals.

Near Rome in 1772, there was found a petrified ox head, of which Father Jacquier has given the description. "The length of the front, taken between the two horns is, he says, two feet and three inches; the distance between the orbits of the eyes is fourteen inches, that from the upper part of the forehead to the orbit of the eye is one foot six inches; the circumference of horn measured in the lower bulge is one foot six inches; the length of a horn measured over all its curvature is four feet; the distance from the tips of the horns is three feet; the interior is a very hard petrifaction: this head was found in a substratum of Pozzolane at a depth of more than twenty feet.*

"There was seen in 1768, in the cathedral of Strasbourg, a very large ox horn, suspended by a chain against a column near the choir; it seemed to me to exceed by three times the normal size of those of our largest oxen. As it is high up, I could not measure the dimensions, but I judged it to be about four-and-a-half feet long, by seven to eight inches in diameter at the large end."†

Lionel Waffer reported that he saw, in Mexico, bones and teeth of a prodigious size: among others a tooth three inches across and four inches long, which the most astute people in the country had examined, judging that the head could not be less than one aune in size. Waffer, *Voyage en Amérique*, 367.

This is perhaps the same tooth of which Father Acosta speaks. "I have seen," he says, "a molar tooth that has greatly astonished me by its enormous size, because it was as big as the fist of a man." Father Torquemado, Franciscan, says that he has in his possession a molar tooth twice as large as a fist and that weighs more than two pounds; he adds that in the same village of Mexico, in the convent of Saint-Augustin, he saw a femur bone so large that the individual to which this bone belonged must have been eleven to twelve cubits high, that is to say, seventeen or eighteen feet, and that the head from which the tooth had been taken was as big as one of the large pitchers that one uses in Castille to keep wine in.

Philippe Hernandés reports that there are found in Tezcaco and Tosuca many bones of extraordinary size, and that among these bones there are molar teeth five inches across and ten inches high; from this one can conjecture that the size of the head to which they belonged was so enormous that two men would have difficulty embracing it. Don Lorenzo Baturini Benaduci also says that in Spanish America, above all in the heights of Santa Fé and in the territory of the Puebla and of Tlascallan, there are found enormous bones and molar teeth, of which one kept in his collection is a hundred times larger than the largest human teeth. *Gigantologie espagnole*, by Father Torrubia, *Journal étranger*, November 1760.

The author of this *Gigantologie espagnole* attributes these enormous teeth and large bones to giants of the human species, but is it credible that there were ever men of whom the head was eight to ten feet in circumference? Is it not astonishing enough that in species of hippopotamus or elephant there were some of this size? We thus think that these enormous teeth are of the same species that were newly discovered in Canada by the Ohio River, which we have said belong to an unknown animal of which the species formerly existed in Tartary, in Siberia, in Canada, and which ranged from Illinois as far as Mexico. And as these Spanish authors do not say that elephant tusks were found together with these large molar teeth in New Spain, that makes us presume that there was in effect a different species from that of the elephant to which these large molar teeth belonged, and that came as far as Mexico. Furthermore, these large hippopotamus teeth seem to have been known in olden times, as St. Augustine says that he saw a molar tooth so large that dividing it would make a hundred molar teeth of an ordinary man (lib. XV, *De civi-*

* *Gazette de France*, 25 September 1772, article on Rome.

† Note communicated to M. de Buffon by M. Grignan, 24 September 1777.

tate Dei, chap. 9). Fulgose also says that, in Sicily, teeth have been found that weigh three pounds each (lib. I, chap. 6).

M. John Sommer reports finding at Chartham, near Canterbury, at seventeen feet in depth, some strange and monstrous bones, some entire, others broken, and four healthy and perfect teeth, each weighing a little more than half a pound, about as large as the fist of a man, all four being molar teeth resembling those of a man closely enough, except for their size. He says that Louis Vives speaks of an even larger tooth (*dens molaris pugno major*), which was shown to him as a tooth of Saint Christopher; he also says that Acosta reports to have seen in the Indies a similar tooth that was pulled out of the earth with several other bones which, assembled and arranged, represent a man of stature that was prodigious, or rather monstrous (*deformed Higness or greatess*). We could have—said M. Sommer judiciously—considered the same of the teeth taken from the earth by Canterbury, if there were not found, with these same teeth, bones which could not be human bones: several people who saw them judged that the teeth and the bones were those of a hippopotamus. Two of these teeth are engraved on a plate that is at the head of No. 272 of the *Philosophical Transactions*, fig 9.

One can conclude from these facts, that most of the large bones found in the heart of the Earth are the bones of elephants and hippopotami; but it seems to me certain, from the immediate comparison of the enormous teeth with rounded points with the teeth of elephant and of hippopotamus, that they belong to an animal much larger than the one and the other, and that today the species of this prodigious animal no longer exists.

Among elephants living today, it is extremely rare to find one where the tusks are six feet long. The largest are commonly five feet to five-and-a-half feet, and consequently the ancient elephant to which belonged the tusk of ten feet in length, of which we possess the fragments, was a giant of this kind as much as that of which we have a femur a third wider and longer than the femurs of ordinary elephants.

It is the same with the species of hippopotamus: I had the two largest molar teeth removed from the largest hippopotamus head that we have in the Cabinet du Roi: one of these teeth weighs ten ounces, and the other nine-and-a-half ounces. I then weighed two teeth, one found in Siberia and another in Canada; the first weighs two pounds twelve ounces, and the second two pounds two ounces. These ancient hippopotami were, as one sees, quite gigantic in comparison with those that exist today.

The example that we have cited of the enormous petrified head of an ox, found near Rome, also proves that there were prodigious giants of this species, and we can show this by several other monuments. We have in the Cabinet du Roi, (1) A horn of a nice greenish color, very smooth and well twisted, which is evidently the horn of an ox; it is twenty-five inches in circumference at the base, and its length is forty-two inches; its cavity contains eleven-and-a-quarter Paris pints. (2) A bone from the interior of the horn of an ox, weighing seven pounds; while the largest bone of our oxen that supports the horn does not weigh more than one pound. This bone was given to the Cabinet du Roi by M. le comte de Tressan, who combines with his taste and his talents much knowledge of natural history. (3) Two bones from the interior of horns of an ox joined by a fragment of skull, which was found at twenty-five feet of depth, in beds of peat, between Amiens and Abbeville, and which was sent to me for the Cabinet du Roi: this fragment weighs seventeen pounds; hence each bone of the horn, separated by the portion of skull, weighs at least seven-and-a-half pounds. I have compared the dimensions and the weights of different bones: that of the largest ox that one could find in the Paris butchery is only thirteen inches long by seven inches in circumference at the base; while of two others, pulled from the heart of the Earth, one is twenty-four inches in length and twelve inches in circumference at the base, and the other twenty-seven inches long and thirteen in circumference. So here there is more than is necessary to show that among the species of ox, as in those of hippopotamus and of elephant, there were once prodigious giants.

4. *We have monuments pulled from the heart of the Earth, and particularly from the bottom of coal and slate mines, that show us that some of the fish and the plants that these materials contain are not species that are alive today.*

On this we will observe, with M. Lehman, that imprints of plants are hardly ever found in slate mines, with the exception of those that accompany coal seams and, by contrast, one generally finds imprints of fish only in copper-bearing slates. Tome III, 407.

It has been observed that the beds of slate charged with petrified fish, in the county of Mansfield, are overlain by a bed of rock termed *stinking*; it is a type of grey slate, which had its origin in stagnant water, in which the fish decayed before being petrified. Leeberoth, *Journal Œconomique*, July 1752.

M. Hoffman, in talking of slates, said that not only were the fish found in it once living creatures, but that the beds of slate were only the deposit of muddy water, which after having fermented and being petrified, was precipitated as very thin beds.

The slates of Angers, says M. Guettard, sometimes show imprints of plants and of fish, which merit attention all the more, in that the plants to which these imprints belong were seaweed, and that those of the fish represent different crustaceans or animals of the class of crayfish, of which the imprints are less common that those of fish or of shells. He adds that after having consulted several authors who have written about fish, crayfish, and crabs, he found nothing resembling the imprints in question, if it is not the sea louse, which is in several reports, but which nevertheless differs by the number of segments, which are thirteen in number; in comparison with which the segments number only seven or eight in the imprints in the slate. The imprints of fish are commonly found strewn with pyritic and white material. One singular fact, which does not regard so much the slate mines of Angers as those of other countries, is the frequency of imprints of fish and the scarcity of shells in these slates, while they are so common in ordinary limestones. *Mémoires de l'Académie des Sciences*, 1757, 52.

One can give demonstrable proofs that all of the coals are composed only of plant debris, mixed with bitumen and sulfur, or rather with vitriolic acid, which can be sensed during combustion. Plants are often seen in large amount in the beds above the seams of coal; and as one successively descends, one sees the nuances of decomposition in these same plants. There are types of coal that are only fossil trees: those found at Saint-Agnès, near Lons-le-Saunier, perfectly resemble logs or trunks of fir: there are clearly seen in them the lines of each annual growth, and also the heart. These types of coal differ from ordinary pines only in that they are oval along their length, and that their rings form so many concentric ellipses. These logs are only barely a foot around, and their bark is very thick and strongly crevassed, like those of old firs, while ordinary firs of similar size always have a fairly smooth bark.

"I have found," says M. de Gensanne, "several seams of this same coal in the diocese of Montpelier: here the trunks are very large, their tissue is very similar to that of chestnut trees three or four feet around. These kinds of fossils give in burning only a faint odor of asphalt; they burn, give a flame, and form cinders like wood. It is this that one commonly calls in France *houille*; it is found very close to the surface of the ground; these houilles usually indicate that there is true earth coal at greater depths." *Histoire naturelle de Languedoc*, by M. de Gensanne, tome I, 20.

These ligneous coals must be regarded as trees deposited in bituminous earth, to which they owe their quality as fossil coals; one never finds them except in these kinds of earth and always fairly close to the surface of the ground. It is not even rare that they form the head of seams of a true coal, and there are some that, having received only a little bituminous substance, have preserved their nuances of woody colors. "I have found some of this species, says M. de Gensanne, "at Cazarets near Saint-Jean-de-Cucul, four leagues from Montpelier; but generally the fracture of this fossil shows a smooth surface, entirely similar to that of jet. There is, in the same canton near d'Aseras, fossil wood that is in part changed into a true white ferruginous pyrite. The mineral matter occupies the heart of the wood, and one notices there very distinctly the ligneous substance, corroded in some way and dissolved by the mineralizing acid." *Hist. nat. de Languedoc*, tome I, 54.

I must admit that I am surprised to see that, following these similar proofs reported by M. de Gensanne himself, who is, besides, a good mineralogist, he nevertheless attributes the origin of coal to clay more or less impregnated with bitumen; not only the facts that I have just cited from him conflict with this opinion, but one will see by these facts that I will put forward that one can attribute the entire mass of all kinds of coals only to plant detritus mixed with bitumens.

I clearly feel that M. de Gensanne does not consider these fossil trees, any more than the peat and even the houille, as true and entirely formed coals, and in this I share his opinion; that which was found by Lons-le-Saunier has been recently examined by M. the President of Ruffey, learned academician of Dijon. He says that these fossil trees closely approach coals in nature, but they are found two or three feet below the surface of the ground across an extent of two leagues and a thickness of three to four feet, and that one can easily recognize species of oak, hornbeam, beech, aspen: that there are logs and bundles of wood, that the bark of the logs is well preserved, that one distinguishes circles of sap and axe marks, and that at a little way away one sees piles of wood chips; that furthermore the coal into which the wood is changed is excellent for welding iron, that nevertheless it gives off a fetid odor when one burns it, from which one can extract alum. *Mémoires de l'Académie de Dijon*, tome I, 47.

"Near the village called Beichlitz, about a league from the town of Halle, two beds composed of bituminous earth and of fossil wood are exploited (there are several mines of this kind in the country around Hesse) and this is similar to that found in the village of Saint-Agnès in Franche-Comté, two leagues from Lons-le-Saunier. This mine is in the terrain of Saxony; the first bed is three-and-a-half toises in perpendicular depth, and eight to nine feet thick. To get to it one passes through a white sand, then a grey and white clay, which serves for roofing and which is three feet thick; one finds still below a good thickness, as much sand as clay, which covers the second bed, which is solely three-and-a-half to four feet thick; one has dug much further down without finding others. These beds are horizontal, but they rise and fall a little like the other known beds. They consist of a brown bituminous earth that is friable when it is dry and resembles decayed wood. Pieces of wood of all sizes are found there, which need to be cut by blows of an axe when they are drawn from the mine while they are still wet. This wood when dry breaks very easily. It is shiny along fractures like bitumen, but one can see there all the organization of wood. It is less abundant than the earth; the workmen put it to one side for their own use.

"A bushel or two quintals of bituminous earth sells for eighteen to twenty French sous. There is pyrite in these beds; its matter is vitriolic; it effloresces and whitens in air; but the bituminous material is not greatly in demand, giving only a feeble heat." *Voyages métallurgiques de M. Jars*, 320ff.

All this proves that in effect this type of fossil wood deposit, which is found so close to the surface of the ground, would be much newer than the seams of ordinary coal, which nearly all plunge deeply into the ground; but this does not stop these ancient seams from being formed of plant debris because, even in the deepest, one can recognize ligneous substances and several other characters that belong only to plants. Besides, there are several examples of fossil wood found in large masses and in very extensive beds, below layers of sandstone and below calcareous rocks. See what I have said about this in this volume, in the article on "Additions on subterranean wood." There is thus no other difference between the true coal and these coalified woods, with more or less decomposition, and also more or less impregnation by bitumens; but the basis for their substance is the same, and both equally owe their origin to plant remains.

M. le Monnier, the first general attending physician of the king and a learned botanist, has found in shale or false slate, which traverses a mass of coal in Auvergne, the impressions of several species of fern that were almost all unknown to him; he thought solely to have noticed the impressions of leaves of the royal fern, of which he has never seen a single foot in all of the Auvergne. *Observations d'Hist. nat. par M. le Monnier*, Paris, 1739, 193.

One would wish that our botanists made exact observations on the plant impressions that are found with these coals in the slates and the shales. One would even need to draw and engrave these plant impressions just as well as those of the crustaceans, the shells, and the fish that these seams enclose, because it will be only after this work that one will be able to pronounce on the present or past existence of all these species, and even on their relative age. All that we know of them today is that there are more unknown forms than others, and in those that one has wished to assign to well-known species, there have always been found differences that are great enough to make us not entirely satisfied with the comparison.

5. *We can show by experiments that are easy to repeat that the glass and powdered sandstone change in a short while to clay by being immersed in water.*

"I placed two pounds of powdered sandstone into a pottery vessel," said M. Nadault, "I filled the vessel with distilled fountain water, so that it covered the sandstone by about three or four fingers of height; I then shook this sandstone for a few minutes and exposed the vessel to the air: a few days later, I saw that there had formed, on the sandstone, a layer more than a quarter-of-an inch thick of very fine yellow earth, very greasy and very ductile. Then, by tilting I poured out the surface water into another vessel, and this earth, being lighter than the sandstone, was separated without being mixed with it: the quantity that I had drawn off by this first washing was too considerable to be able to think that, in a space of time as short as this, there could have been produced a fairly great decomposition of the sandstone, to have produced so much earth. I therefore judged that this earth must already have been in the sandstone in the same state in which I had drawn it off, and that there would perhaps have been thus a continual decomposition of the sandstone in its own bed. I therefore refilled the vessel with new distilled water; I agitated the sandstone for a few instants, and three days later, I found again on the sandstone a layer of earth of the same quality as the first, but thinner by half. Having put to one side these types of secretions, I continued, over the course of more than a year, this same operation, and these experiments that I had commenced in the month of April; and the quantity of earth that the sandstone produced diminished little by little, so that at the end of two months, on transferring the water from the vessel that contained it, I no longer found on the sandstone more than a pellicle of earth, which was not even a line in thickness. But, also for the rest of the year, and as long as the sandstone was in the water, this pellicle never failed to form in the space of two or three days, without increasing or diminishing in thickness, with the exception of the time when I was obliged, because of the frost, to put the vessel under cover, when it seemed to me that the decomposition of the sandstone took place a little more slowly. Some time after having put this sandstone into the water, I noticed a great quantity of bright silvery flecks, like those of talc, which were not there before, and I judged that this was here the first state of decomposition; that these molecules, formed of many tiny layers, were exfoliating, as I have seen happening to glass in some circumstances, and that these flecks were then attenuated little by little in the water, until they became so small that they no longer had surface enough to reflect light, and they took on the form and the properties of a veritable earth. I thus collected and set aside all the terrigenous secretions that these two pounds of sandstone produced over the course of more than a year; and when this earth was thoroughly dried, it weighed about five ounces. I also weighed the sandstone after having dried it, and it had diminished in weight by the same proportion, such that a little more than one-sixth part of it had become decomposed; all this earth was however of the same quality, and the last secretions were as greasy, as ductile, as the first, and always of a yellow approaching orange; but as I still saw there some brilliant flecks, a few molecules of sandstone, which were not entirely decomposed, I put this earth with water into a glass jar, and I left it exposed to the air, without stirring it, for all of a summer, adding from time to time a little new water to replace that evaporated: a month later, this water started to become corrupted, and it became greenish, with unpleasant odor. The earth also seemed to be in a state of fermentation or of putrefaction, because a large number of air bubbles rose from it; and although it superficially retained its yellowish color, that which was at the bottom of the vessel became brown, and this color extended from day to day, and seemed to become darker; with the result that at the end of summer, this earth had become absolutely black. I let the water evaporate without putting any more into the vessel, and having taken out the earth, which resembled the grey clay closely enough except for being humectified, I dried it with the heat of a fire, and while it was being heated, it seemed to me that it exhaled a sulfurous odor. What surprised me more was that as it dried the black color became a little effaced, and it became as white as the whitest clay; from which, one can conjecture, that it was consequently a volatile material, which gave it the brown color. Acid spirits made no impression on this earth; and having subjected it to a fairly intense heat, it did not redden at all as does grey clay, but it kept its whiteness. From this, it seems to me evident that this material that the sandstone produced for me, in attenuating and being deposited with water, is a true white clay." Note communicated to M. de Buffon by M. Nadault, correspondent of the Academy of Sciences, former advocate general of the Chamber of Accountants of Dijon.

6. *The movement of the waters from the east to the west has worked the surface of the Earth in this sense: in all of the continents of the world, the slope is steeper from the coast of the west than from the coast of the east.*

This is seen in the continent of America, of which the slopes are extremely steep toward the seas of the west, and of which all of the terrain extends in a gentle slope and ends almost everywhere in the great plains by the coast of the sea to the east. In Europe, the line of the summit of Great Britain, which extends from the north to the south, is much closer to the western coast than the eastern coast of the ocean; and for the same reason, the seas that are to the west of Ireland and of England are deeper than the sea that separates England and Holland. The line from the summit of Norway is much closer to the ocean than to the Baltic Sea; the mountains of the general summit of Europe are much higher toward the west than toward the east; and if one takes a part of this summit from Switzerland up to Siberia, it is much closer to the Baltic Sea and the White Sea than it is to the Black Sea and the Caspian Sea. The Alps and the Apennines rise much closer to the Mediterranean than to the Adriatic Sea. The mountain chain that emerges from the Tyrol and that extends into Dalmatia and as far as the point of the Peloponnese hugs, so to speak, the Adriatic Sea, while the eastern coasts on the opposing side are lower. If, in Asia, one follows the chain that extends from the Dardanelles to the straits of Bab-el-Mandeb, one finds that the summits of Mount Taurus, of Lebanon, and of all Arabia hug the Mediterranean and the Red Sea; and in the east, there are vast continents where flow the rivers with long courses, which flow into the Persian Gulf. The summit of the famous mountains of Ghats approaches the western seas more closely than the eastern seas. The summit that extends from the western frontiers of China to the point of Malacca is yet closer to the western sea than to the eastern sea. In Africa, the chain of Mount Atlas sends, into the Canary seas, rivers that are less long than those it sends into the interior of the continent, and which flow far, to become lost in lakes and great swamps. The high mountains that are in the west toward Cape Verde and in all of Guinea, having turned around to Congo, arrive at the Mountains of the Moon, and stretch out toward the Cape of Good Hope, occupying regularly enough the interior of Africa. One nevertheless recognizes, in considering the sea to the east and to the west, that that of the east is not so deep, with a great number of islands, and hence the deepest part of the western sea is much closer to this chain than the deepest of the eastern or the Indian seas.

One thus sees in general, in all the great continents, that the points of division are always much closer to the seas to the west than to the seas to the east; that the facing sides of the continents are always toward the east, and shortened to the west; that the seas of the western coasts are deeper and strewn with many fewer islands than the eastern ones; and one will even recognize that in all these seas, the coasts of the islands are always higher and the seas that bathe them deeper to the west than to the east.

Notes on the Fifth Epoch

1. *There are animals and even men so brutish that they prefer to languish in their thankless native land rather than taking the trouble to settle more comfortably elsewhere.*

I can cite a striking example of this: the Maillés, a small uncivilized nation of Guyana, a small distance from the mouth of the Ouassa River, have no home other than the trees, above which they stay all the year, because their land is always more or less covered by water: they come down from the trees only to go into canoes to look for their food. Here is a singular example of stupid attachment to the land of their birth; because for these savages to go and live on the land like others, they would distance themselves by only a few leagues from the drowned savannahs where they were born and where they wish to die. This fact, cited by a few travelers,* was confirmed to me by several witnesses who

* The Maillés, one of the uncivilized nations of Guyana, live along the coast, and as their country is often drowned, they have constructed their huts upon the trees, at the foot of which they keep the canoes with which they go to seek what is necessary for them to live. *Voyage de Demarchais*, tome IV, 352.

recently saw this little nation, composed of three or four hundred savages; they keep themselves in effect to the trees above the water, and live there for all of the year; their terrain is a great sheet of water during the eight or nine months of rain; and during the four months of summer, the land is only a filthy mud, upon which there forms a small crust, five or six inches in thickness, composed of grass rather than of earth, and beneath which there is a great thickness of foul and tainted water.

Notes on the Sixth Epoch

1. *The Caspian Sea was formerly much larger than it is today. This supposition is well founded.*

"In traveling over," said M. Pallas, "the immense deserts that extend between the Volga, the Yaik, the Caspian Sea, and the Don, I noticed that these steppes or sandy deserts are in all parts surrounded by a raised shore, which embraces a large part of the bed of the Yaik, the Volga, and the Don, and that these very deep rivers, before they penetrated into this enclosure, are full of islands and shallows as soon as they begin to fall into the steppes, where the great river of Kuman goes to lose itself in the sands. From these combined observations, I conclude that *the Caspian Sea once covered all these deserts*; that it did not formerly have other boundaries than these same raised shores that surround them in all parts, and that it communicated by means of the Don with the Black Sea, supposing even that this sea, like that of Azov, was not part of it."*

M. Pallas is without doubt one of our most learned naturalists, and it is with the greatest satisfaction that I see him here entirely sharing my view on the ancient extent of the Caspian Sea, and on the well-founded probability that it once communicated with the Black Sea.

2. *Tradition has conserved for us only a memory of the submergence of the Taprobane . . . There were greater and more frequent upheavals in the Indian Ocean than in any other part of the world.*

The most ancient tradition that remains of these founderings in the lands of the south is that of the loss of Ceylon, of which one believes that the Maldives and the Laquedives were once part. These islands, as well as the banks and reefs that reign from Madagascar as far as the point of India, seem to indicate the summits of the lands that united Africa with Asia; because these islands are almost all on the northern side, lands and banks, which project very far under the waters.

It would also seem that the islands of Madagascar and of Ceylon were formerly united with the continents that neighbor them. These separations, and these great upheavals in the seas of the south, were produced for the most part by the collapse of caverns, by earthquakes, and by the explosion of subterranean fires. But, there were also many lands invaded by the slow and progressive movement of the sea from the east to the west; the parts of the world where this effect is most noticeable are the regions of Japan, of China, and of all the eastern parts of Asia. Those seas situated to the west of China and of Japan are only accidental, so to speak, and perhaps yet more recent than our Mediterranean.

The Sunda islands, the Mollucans, and the Philippines are only upheaved lands, and are still full of volcanoes. There are many of these also in the islands of Japan, and people say that this is the place in the universe that is most subject to earthquakes; one finds

* *Journal historique et politique*, November 1773, article on « Pétersbourg. »

there quantities of fountains of hot water. Most of the other islands of the Indian Ocean show little other than peaks of summits of isolated mountains that vomit fire. The Isle of France and the Isle of Bourbon show two of these summits, almost entirely covered in materials thrown out by the volcanoes; these two islands were uninhabited when this discovery was made.

3. *In Guyana, the rivers are so close one to another, and at the same time so swollen, so fast-flowing in the rainy season, that they entrain immense amounts of silt, which are deposited over all the lowlands and on the bottom of the sea as muddy sediments.*

The coasts of French Guyana are so low that they are rather beaches all covered in mud on a very gentle slope, which begin on land and extend to the bottom of the sea a very great distance away. The great ships cannot approach the Cayenne River without grounding, and warships are obliged to remain two or three leagues out to sea. These muds on a gentle slope extend all along the coast, from Cayenne as far as the Amazon River: along all that great extent there is only found mud, with no sand at all, and all of the fringes of the sea are covered with mangroves. But, seven or eight leagues above Cayenne, from the northwest coast as far as the Maroni River, there are a few coves of which the bottoms are sandy and with rocks that form breakwaters. However, these are mostly covered by mud, as much as by beds of sand; and this mud becomes thicker with distance from the edge of the sea: the little rocks do not stop this terrain from being a very gentle slope to several leagues' extent into the land. That part of Guyana to the northwest of Cayenne forms higher land than those to the southeast. One has demonstrable proof of that; because, all along the edge of the sea, there are great drowned savannahs, which border the coast, and of which most are dried out in the northwest, while they are all covered with seawater in the southeast parts. Besides these lands that are now drowned by the sea, there are others farther away, and which were similarly once drowned; one finds also fresh water in a few places of the savannahs, but these do not produce any mangroves, but only many latanier palms. One does not find even a single stone on all of these low coasts; the tide does not stop from rising six or seven feet in height, while the currents are opposed to it, because these are all directed toward the Antilles islands. The tide is very noticeable when the rivers are low, and one can see this as far as forty or even fifty leagues into these rivers; but in winter, that is to say, in the rainy season, when the rivers are swollen, the tides are barely noticeable over one to two leagues, as the current in these rivers is so rapid, and it assumes the greatest impetuosity during the hour of ebb.

The large sea tortoises come to lay their eggs at the bottoms of these coves of sand, and they are never seen frequenting muddy terrain; hence from Cayenne as far as the Amazon River, there are no sea tortoises, and one goes to fish for them from the Caura River as far as the Maroni River. It seems that the mud is gaining ground every day over the sand, and with time, this northwest coast of the Cayenne will be covered with it as is the southeast coast; because the sea tortoises, which only wish for sand in which to lay their eggs, are slowly moving away from the Caura River, and for several years, one has been obliged to go to look for them farther, by the coast of the Maroni River, where the sands have not yet been covered.

Beyond the savannahs, of which some are dry and others are drowned, extends a cordon of hills, which are all covered with a great thickness of earth, planted everywhere with old forests. Generally these hills are 350 to 400 feet in height, but at a greater distance, one can find higher hills, maybe twice as high, on moving as far as ten or twelve leagues into these lands: most of these mountains are evidently old extinct volcanoes. There is however one called *la Gabrielle*, at the summit of which is found a large pool or small lake, which sustains caymans in a fair number, of which apparently the species has been conserved since the time that the sea covered this hill.

Beyond this Gabrielle Mountain, there are only small valleys, knolls, swamps, and volcanic materials, which are not at all in large masses, but are broken into small blocks; the most common stone, of which the waters have carried blocks as far as Cayenne, is that called cockroach stone which, as we have said, is not at all a stone, but a lava from a volcano: it has been called a cockroach stone because it has holes and the insects called cockroaches live in the holes of this lava.

4. *The race of giants in the human species has been destroyed for many centuries in the place of its origin in Asia.*

One cannot doubt that there have been individual giants in all the climes of the Earth, since in our times, they can be seen to be born in all countries, and recently, one was seen to be born within the confines of Lapland, on the coast of Finland. But one is not equally sure that there used to be constant races of them, and still less that there were entire peoples of giants; nevertheless, the witness of several ancient writers, and that of the Holy Scripture, which is yet more ancient, seems to indicate to me clearly enough that there were races of giants in Asia; and we think it necessary to show here the most positive passages on this subject. It is said, Numbers XIII, verse 34: *We have seen the giants of the race of Hanak, in the eyes of whom we cannot seem any larger than grasshoppers.* And in another version, it is said: *We have seen the monsters of the race of Enac, next to which we are not larger than locusts.* Although this has the air of exaggeration, common enough in the oriental style, this nevertheless proves that these giants were very large.

In Deuteronomy, chapter XXI, verse 20, there is mention of a very large man *of the race of Arapha, who had six toes on his feet and his hands.* And one sees in verse 18, that this race of Arapha was *of the gigantic kind.*

One finds again in Deuteronomy several passages that prove the existence of giants and of their destruction: *A numerous people*, it is said, *of a great height, like those of Énacim, who the Lord destroyed*; chapter II, verse 21. And it is said, verses 19 and 20: *The country of Ammon is reputed as a country of giants, in which once lived the giants that the Ammonites called the Zomzommim.*

In Joshua, chapter II, verse 22, it is said: *The sole giants of the race of Énacim, who rested among the children of Israel, were in the towns of Gaza, of Gette, and of Azots; all the other giants of this race were destroyed.*

Philo, St. Cyril, and several other writers seemed to think that the word "giants" indicated only men who were superb and godless, and not men of extraordinary size of body; but this opinion cannot be upheld, since it is often a question of the height and bodily strength of these same peoples.

In the prophet Amos, it is said that people of the Amores were so tall that they were compared to cedars, without giving other measures of their height.

Og, king of Bashan, had a height of nine cubits, and Goliath, of ten cubits and one palm. The bed of Og was nine cubits long, that is to say, thirteen-and-a-half feet, and in width four cubits, which makes six feet.

The corselet of Goliath weighed two hundred eight pounds and four ounces, and the iron of his lance weighed twenty-five pounds.

These witnesses seem to me sufficient, that one can believe with some foundation that there once existed in the continent of Asia, not only individuals but also races of giants, who have been destroyed, and of which the last ones still lived in the times of David; and sometimes Nature, which never loses its rights, seems to climb back to this same point of force of production and of development; because in almost all climes of the Earth, there appear, from time to time, men of extraordinary height, of seven-and-a-half, eight, and even nine feet. Because, independently of well-proven giants, of whom we have made mention, *Supplement*, tome V, we can cite an infinite number of other examples, reported by writers ancient and modern, giants of ten, twelve, fifteen, eighteen feet in height, and even more; but I am strongly persuaded that one needs to greatly reduce these last estimates: elephant bones have often been mistaken for human bones, and besides, Nature, as she is known to us, does not offer us any species with disproportions so great, except perhaps in the species of hippopotamus, of which teeth found in the heart of the Earth are at least four times larger than the teeth of today's hippopotami.

The bones of the alleged King Teutobachus, found in Dauphiné, have been the subject of a dispute between Habicot, surgeon of Paris, and Riolan, Doctor in Medicine and celebrated anatomist. Habicot wrote, in a small work that has for title *Gigantostéologie*, that these bones were in a brick sepulcher eighteen feet below ground, surrounded by fine sand; he gives neither exact description, nor dimensions, nor the number of the bones. He alleges that these bones were really human bones, all the more so, he said, as no animal possesses such. He adds that it is the masons who, working at the house of the lord of Langon, a gentleman of Dauphiné, found, on 11 January 1613, this tomb, near to the ruins of the chateau of Chaumont; that this tomb was of brick, that is was thirty feet long, twelve in width, and eight in depth, counting in this the coverstone, in the middle of which was a grey stone upon which was engraved *Teutobachus rex*; that, upon opening this tomb, one saw a human skeleton twenty-five-and-a-half feet in length, ten in width at the shoulders, and five in thickness; that before touching these bones the head was measured at five feet in length and ten in circumference. (*I must observe that the proportion of the length of the human head to its body is not a fifth, but one-seven-and-a-half part; hence for this head of five feet, the body should be thirty-seven-and-a-half feet tall*). Finally, he said that the lower jaw was six feet around, the orbits of the eyes seven inches around, each clavicle four feet long, and that most of the bones turned into powder after being in contact with air.

The doctor Riolan published in the same year of 1613 an account under the name of *Gigantomachie*, in which he said that the surgeon Habicot gave, in his *Gigantostéologie*, false measures of the size of the body and the bones of the alleged giant Teutobachus; that he, Riolan, measured the bone of the thigh, that of the leg, with the astragal joint at the calcaneum, and that he found only six-and-a-half feet for them, including the pubic bone, which would make thirteen feet instead of twenty-five for the height of the giant.

He then gave reasons that made him doubt that these were human bones, but rather were those of an elephant.

One or two years after the publication of the *Gigantostéologie* of Habicot and of the *Gigantomachie* of Riolan, a brochure appeared, under the title of *Imposture discovered of the supposed human bones, falsely attributed to King Teutobachus*; in which one discovers nothing else, except that these bones are not human bones, but fossil bones generated by virtue of the Earth: and yet another little book, with an author's name, and which it says that in truth there are human bones among these bones, but that there were others there too that were not human. Then in 1618, Riolan published a text, under the name of *Gigantologie*, where he contended that not only were the bones in question not human bones, but further, that men in general were never larger than they are today.

Habicot responded to Riolan in the same year of 1618; and he said that he had offered to King Louis XIII his *Gigantostéologie*, and that in 1613, at the end of July, the bones described in this work were exposed to the eyes of the public, and that these are really human bones: he cites a large number of examples, drawn from ancient and modern writers, to prove that there used to be humans of excessive size: he persists in saying that the calcaneum bone, the tibia, and the femur of the giant Teutobachus were joined to each other, being more than eleven feet in height.

He then cites letters that were written to him at the time of the discovery of these bones, and that seem to confirm the reality of the fact of the tomb and of the bones of the giant Teutobachus. It seems from the letter of the lord of Langon, dated from Saint-Marcellin in Dauphiné, and from another from M. Masurier, surgeon at Beaurepaire, that silver coins were found with the bones. The first letter is conceived in the following terms: "As his Majesty wishes to have the rest of the bones of the King Teutobachus, with the silver coins that were found there, I can tell you beforehand that your contrary advisers are very poorly founded, and that if they knew their profession, they would not doubt that these bones are truly human bones. The doctors of medicine of Montpellier have been brought here, and would well have wished to have these bones for money. M. le Maréchal of Lesdiguières had them taken to Grenoble to be seen there, and the doctors and surgeons of Grenoble recognized them as human bones; and so it is only ignorant people who deny this truth, & etc." *Signed*, Langon.

Furthermore, in this dispute, Riolan and Habicot, one a doctor and the other a surgeon, exchanged more insults than they wrote down facts and reasons. Neither one nor the other had enough sense to describe exactly the bones in question; but both, transported by pride and partiality, made descriptions in a manner to remove all confidence. It is thus very difficult to pronounce definitively upon the species of these bones; but if they were in effect found in a brick tomb, with a coverstone upon which was written *Teutobachus rex*; if coins were found in this tomb; if it contained a single cadaver twenty-four or twenty-five feet in length; if the letter of the lord of Langon contains truth, one could hardly doubt the essential fact, that is to say, of the existence of a giant twenty-four feet in height, if one is not to suppose an extraordinary concurrence of mendacious circumstances. But also, the fact has not been proven in a positive enough manner, in order that one cannot greatly suspect it. It is true that several writers, moreover worthy of belief, have talked of giants that are as large and yet larger. Pliny reported that during an earthquake in Crete, a mountain having opened up, there was there found a body sixteen cubits long, which some said was the body of Otus, and other that of Orion. The sixteen cubits give twenty-four feet in length, that is to say, the same as that of King Teutobachus.

One finds in the memoir of M. le Cat, academician of Rouen, an enumeration of several giants of excessive size; that is to say, two giants of which the skeletons were found by the Athenians near their town, one of thirty-six and the other thirty-four feet in height; another of thirty feet was found in Sicily near Palermo, in 1548; another of thirty-three feet, also found in Sicily in 1550; and yet another, similarly found in Sicily near Mazarino, who was 30 feet tall.

Despite all these witnesses, I think one will have much difficulty persuading oneself that there ever existed men of thirty or thirty-six feet in height; it would be already too much not to refuse to believe that there were some of twenty-four; nevertheless these sightings multiply, become more positive, and go so to speak by nuances of growth as one descends. M. le Cat reports that in 1705, near the banks of the Morderi River, at the foot of the mountain of Crussol, the skeleton of a giant twenty-two-and-a-half feet high was found; and that the Dominicans of Valence possess one part of his leg, with the articulation of the knee.

Platerus, the celebrated doctor, testified to have seen at Lucerne the skeleton of a man at least nineteen feet tall.

The giant Ferragus, killed by Roland, nephew of Charlemagne, was eighteen feet high.

In the sepulchral caverns of the island of Tenerife, the skeleton of a Guanche was found that was 15 feet tall, and of which the head had eighty teeth. These three facts are reported, like the preceding ones, in the memoir of M. le Cat upon giants. He cites also a skeleton found in a ditch near the convent of the Dominicans of Rouen, of which the skull held a bushel of wheat, and of which the bone of the leg was around four feet long, which gives seventeen or eighteen feet for the height of the entire body. On the tomb of this giant there was an engraved inscription, where one reads: *Here lies the noble and powerful lord the Chevalier Ricon de Valmont and his bones.*

One finds in the *Journal littéraire* of Abbé Nazari that in high Calabria, in the month of June 1665, there was unearthed in the gardens of the lord of Tiviolo, a skeleton eighteen Roman feet in length; that the head was two-and-a-half feet long; that each molar tooth weighed about one-and-one-third ounces, and that this skeleton was resting on a mass of bitumen.

Hector Boëtius, in his *History of Scotland*, book VII, reports that there are still preserved a few bones of a man, named in contra-verity *Little John*, who was thought to have been fourteen feet tall (that is to say, thirteen feet, two inches and six lines of France).

There is found in the *Journal of Savants*, year 1692, a letter from P. Gentil, priest of the oratory, professor of philosophy at Angers, where he says that having had news of a discovery made of a gigantic cadaver in the borough of Lassé, nine leagues from this town, he himself went to this place to inquire about this fact. He learned that the local priest, having had his garden dug, there was discovered a sepulcher that enclosed a body seventeen feet, two inches long, which no longer had skin. This cadaver had other bodies between its arms and legs, which could have been its children. There was found at the same place fourteen or fifteen other sepulchers, some of ten feet, others of twelve, and others even of fourteen feet, which enclosed bodies of the same length. The sepulcher of this giant stayed exposed to the air for more than a year; but as this attracted too many visits to the priest, he had it covered with earth and planted three trees over it. These sepulchers were of a stone similar to chalk.

Thomas Molineux saw, at the School of Medicine at Leyden, a prodigious human frontal bone; its height, taken from its junction from the bone of the nose to the sagittal suture, was nine-and-one-twelfth inches, its width twelve-and-two-tenth inches, its thickness half an inch, that is to say, each of these dimensions was double that of the corresponding dimension of the frontal bone as it is in men of normal stature; and hence the man to whom this gigantic bone belonged was probably as large again as ordinary men, that is to say, that he was eleven feet tall. This bone was very certainly a human frontal bone; and it seems that it did not acquire this volume through a morbid illness; because its thickness was proportional to its other dimensions, which is not the case in bones with defects.

In the collection of M. Witreu in Amsterdam, M. Klein says he saw a frontal bone, from which it seemed to him that the man to whom it belonged was thirteen feet and four inches in height, that is to say, about twelve-and-a-half feet in France.

After all these facts I have just outlined, and those that I have discussed here before on the subject of the Patagonians, I leave the readers in the same embarrassment in which I find myself, to be able to pronounce upon the real existence of these giants of twenty-four feet; I cannot persuade myself that at any time and by any means or any circumstance, the human body could raise itself to dimensions so enormous; but I believe at the same time that one can hardly doubt that there were giants ten, twelve, and perhaps fifteen feet in height; and that it is almost certain that in the first ages of living Nature, there existed not only gigantic individuals in great numbers, but even a few constant and successive races of giants, of which that of the Patagonians is the only one that was conserved.

5. *One finds above the Alps an immense and nearly continuous extent of valleys, of plains, and of mountains of ice, & etc.*

Here is what M. Grouner and a few other good observers and eyewitnesses report on this subject.

In the highest regions of the Alps, the waters, being sourced annually from the fount of the snows, freeze upon all of the aspects and at all points on these mountains, from their bases up to their summits, especially in the vales and on the slopes of those that are clustered. Hence, the waters have, in these valleys, formed mountains that have rocks at their core, and other mountains that are entirely of ice, which are six, seven, to eight leagues of extent in length, by a league in width, and often a thousand to twelve hundred toises in height; they join the other mountains at their summit. These enormous piles of ice grow in extent on prolonging themselves in the valleys; thus it is clear that all the glaciers are growing progressively; although in warm and rainy years, not only is their progress stopped, but even their immense mass diminishes . . .

The height of freezing, fixed at 2,440 toises at the equator, for the high isolated mountains, is not at all a rule for groups of mountains that are frozen from their base to their summit; they never thaw. In the Alps, the height of degree of freezing is fixed at 1,500 toises of elevation, and all the parts below that height thaw entirely; while those that are heaped up freeze at a lower height and never thaw at any point of their elevation from their base, as the degree of cold is so augmented by the masses of frozen matter united in the same space.

All the glacial mountains of Switzerland, together, occupy an extent of sixty-six leagues from east to west, measured in a straight line, from the western boundary of the canton of Valais toward the Savoy, as far as the eastern borders of the canton of Bendner toward the Tyrol. This forms an interrupted chain, of which several arms extend from south to north upon a length of about thirty-six leagues. The great Gotthard, the Furka, and the Grimsel are the highest mountains of this part. They occupy the center of these chains that divide Switzerland into two parts; they are always covered with snow and with ice, which has made them be given the generic name of *glaciers*.

One divides these glaciers into glacial mountains, vales of ice, fields of ice, or glacial seas, and into *gletschers* or piles of ice blocks.

The glacial mountains are these large masses of rocks, which rise up to the sky, and which are always covered with snow and with ice.

The vales of ice are hollows that are much more elevated between the mountains than the lower vales; they are always full of snow, which accumulates there and forms heaps of ice, which are often several leagues in extent, and which join with the high mountains.

The fields of ice or glacial seas are terrains of gentle slope, which occur around the mountains. They cannot be called vales, because they are not deep enough; they are covered in thick snow. These fields receive water from the source of snows that descends from the mountains and refreezes. The surface of this ice alternately melts and freezes, and all these places are covered with thick layers of snow and of ice.

The gletschers are heaps of ice blocks formed by ice and snow that tumbled down the mountains. These snows refreeze and pile up in different manners; this means that one can divide them into gletschers in hills, in linings, and in walls of ice.

The hills of ice rise between the summits of the high mountains: they themselves have the form of mountains, but no rock at all comes into their structure; they are composed entirely of pure ice, which is sometimes several leagues in length, a league in width, and half a league of thickness.

The linings of ice blocks are formed in the higher valleys and on the sides of the mountains that are covered as with draperies of ice shaped into points; they pour their superfluous water into the lower valleys.

The walls of ice are scarped linings that terminate the valleys of ice, which have a flattened form, and which appear from a distance like agitated seas, of which the waves have been seized and frozen in the moment of their agitation. These walls are not at all bristling with points of ice; often they form columns, pyramids, and towers that are enormous in height and in width: shaped into several faces, sometimes hexagons, and blue or celandine green in color.

Heaps of snow form also on the sides and at the feet of mountains, which are then irrigated by water from melted snows and covered by fresh snows. One also sees ice blocks that accumulate in masses, and which are not held either in vales or within hills of ice; their position is horizontal or inclined; all of these detached masses are called *beds* or *layers of ice* . . .

The interior heat of the Earth undermines many of these mountains of ice from beneath, and entrains currents of water, which melt their lower surfaces; thus these masses subside gently under their own weight, while their height is maintained by the waters, the snow, and the ice, which progressively cover them. This subsidence often produces horrible creaks; the crevasses that open upon the thickness of the ice form precipices that are as troublesome as they are frequent. These abysses are all the more perfidious and deadly in being normally covered with snow. Travelers, the curious, and hunters, who follow the deer, the chamois, the ibex, or who search for ores of crystal, are often swallowed up in these chasms, and thrown back to the surface by the torrents that rise from the bottom of these abysses.

The soft rains quickly melt the snows; but all the waters produced by this do not fall into the abysses below through the crevasses; a large part refreezes, and falls onto the surface of the ice, augmenting its volume.

The warm winds from the south, which generally prevail in the month of May, are the most powerful agents in destroying the snows and the ice; hence, their melting, announced by the murmur of glacial lakes, and by the horrible fracas from the shock of stones and ice that fall indiscriminately from the heights of the mountains, and carrying, through all parts in the lower valleys, torrents of water, which fall from the rocky crags at more than 1,200 feet in height.

The Sun has only a slight power on the snows and on the ice, to make them melt. Experience has shown that these ice masses formed over a very long period of time, under enormous loads, in a degree of cold so multiplied and with water so pure, that this ice, say I, would be of a material so dense and purged of air that small pieces of ice exposed to the fiercest sun on the plain, over a whole day, melt only with difficulty.

Although the mass of these glaciers melts in part each year during the three months of summer, when the rains, the winds, and the greater warmth in some years halt the progress that the glaciers have made in many other years: nevertheless it is proven that these *glaciers show a constant growth and that they are advancing*; the annals of the country prove this; authentic documents show it, and tradition is invariable on this subject. Independently of these authorities and of journalistic observations, this progression of the glaciers is proven by the *larch forests, which have been engulfed by the ice and of which the tips are still above the surface of the ice*; these are irrefutable witnesses that attest to the progress of the glaciers, just as is the top of the *clock tower of a village* that has been buried beneath the snows, and that one sees only during extraordinary melts. This progression of the glaciers cannot have any cause other than the growth in intensity of the cold, which increases in the glacial mountains because of the masses of ice. And, it is proven that in the glaciers of Switzerland, the cold is today deeper, but not as long-lasting as in Iceland, of which the glaciers, like those of Norway, have much in common with those of Switzerland.

The massif of the glacial mountains of Switzerland is composed like those of all the high mountains; the core is a vitreous rock, which extends up to their summit; the part below, starting from the point where they were covered by the waters of the sea, is composed of a coating of calcareous rocks, like all of the massif of mountains of a lower order, which are grouped around the base of the primitive mountains of these glaciers; finally these calcareous masses have, as a base, shale produced by the deposition of silt from water.

The vitreous masses are crystalline rocks, granites, quartz; their fissures are filled with metals, with half-metals, with mineral substances, and crystals.

The calcinable masses are rocks with lime and marbles of all types of colors and varieties, chalks, gypsum, spars, and alabasters, and so on.

The shaley masses are slates of different qualities and color, which contain plants and fish, and which are commonly sited at fairly considerable heights; their bed is not always horizontal, it is often inclined, and even sinuous and perpendicular in some places.

One cannot call into doubt the ancient sojourn of the waters of the sea upon these mountains that today form glaciers; the immense quantity of shells that can be found there attests to this, as do the shales and other rocks of this type. The shells are distributed there by families, or they are well mixed one with another, and one can find them at very great heights.

There is here leave to think that these mountains did not form continuous glaciers in deep antiquity, not even since the waters of the sea abandoned them, although it seems by their very great distance from the sea, which is a hundred leagues away at its closest, and by their excessive height, that they were the first upon the continent of Europe to emerge from the waters. They had in ancient times their volcanoes; it seems that the last to have become extinct is that of the mountain of Myssenberg, in the canton of Schwyz. These two principal summits, which are very high and isolated, are terminated by cones, as are all the mouths of volcanoes; and one can still see the crater in one of these cones, which is cut to a very great depth.

M. Bourrit, who had the courage to make a great number of forays into the glaciers of the Savoy, says "that one cannot doubt the growth of all the glaciers of the Alps; that the quantity of snow that has fallen there during the winters is greater than the quantity melted during the summers; that not only is the same cause operating, but that the masses of ice already formed must always grow more, because this results in more snow and a lesser melt. Thus there is no doubt that the glaciers will go on to grow, and even in an increasing progression."*

This indefatigable observer has made a great number of journeys among these glaciers, and in talking of the *glatcher*, or the Glacier of Bossons, he says "that it seems to grow each day; that the soil that it occupies presently was, a few years ago, a cultivated field, and that the ice still grows every day."† He reports that "the growth of ice seems demonstrable not only at this place, but in several others; that there is still the memory of communication that existed once between Chamonix and the Val d'Aosta, and that the ice has completely closed it; that the ice in general must be growing in extending first from summit to summit, and then from valley to valley, and that it is thus that there is made communication of the ice of Mont Blanc with that of other mountains and glaciers of the Valais and of Switzerland."‡ It seems, he says besides, that "all the mountain lands were, in ancient times, not as full of snow and ice as they are now . . . One dates back to only a few centuries ago the disasters that took place through the growth of snow and ice, by their accumulation in many valleys, by the collapse of the mountains themselves and of the crags: these are almost continual accidents and this annual growth of ice, which can alone provide reason for that which one knows of the history of this land concerning the people who inhabited it formerly."§

6. *But despite what the Russians have said about this, it is very doubtful that they passed the northern point of Asia.*

M. Engel, who regards as impossible the passage to the northwest through the bays of Hudson and of Baffin, seems by contrast persuaded that one will find a shorter and a more sure passage by the northeast; and he adds to reasons he gives on this, which are weak enough, a passage of M. Gmelin who, talking of attempts made by the Russians to discover this passage to the northeast, said *that the manner by which one has proceeded to these discoveries will make in its time the subject of the greatest astonishment of all the world, when one will have the authentic account of this; and that depends solely*, he adds, *on the noble will of the Empress*. What thus will be, says M. Engel, this subject of astonishment, if it is not to learn that the passage regarded until today as impossible is very practicable? Here is the sole fact, he adds, "which could surprise those whom one has tried to scare, by the stories published with the aim of disheartening the navigators, and so on."¶

I first remark that one needs be sure of things, before making such an imputation to the nation of Russia: secondly, it seems to me to be poorly founded, and the words of M. Gmelin could well signify a completely contrary interpretation to that given them by M. Engel, that is to say, that one will be greatly astonished, as one will know that there exists no practicable passage at all to the northeast. What confirms me in this opinion, independently of the general reasons that I have given, is that the Russians themselves have only recently tried for such discoveries in going up from Kamchatka, and not at all in descending from the point of Asia. The Captains Bering and Tschirikov have, in 1741,

* *Description des glacières de Savoie*, by M. Bourrit, Geneva, 1773, 111–12.
† *Description des aspects du Mont Blanc*, by M. Bourrit, *Lausanne*, 1776, 8.
‡ *Description des aspects du Mont Blanc*, 13, 14.
§ *Description des aspects du Mont Blanc*, 62, 63.
¶ *Histoire générale des voyages*, tome XIX, 415ff.

made a reconnaissance of the parts of the coast of America as far as the fifty-ninth degree, and neither one nor the other came by the sea of the north along the coasts of Asia. Or, to put it better, this proves sufficiently that the passage is not as practicable as M. Engel supposes it to be; otherwise, they would have preferred to send their navigators by this route, rather than make them leave from Kamchatka, to make the discovery of western America.

M. Muller, sent with M. Gmelin by the Empress to Siberia, has a quite different opinion from M. Engel: having compared all of the accounts, M. Muller concluded by saying that there is only a very small separation between Asia and America, and that this strait offers one or several islands, which serve as route or as common stations for the inhabitants of the two continents. I think this opinion well founded, and M. Muller assembles a great number of facts to support it. In the subterranean dwellings of the inhabitants of the island of Karaga, one sees beams made of large fir trees, which this island produces none of, any more than do the lands of Kamchatka, which it closely neighbors: the inhabitants say that this wood comes to them by a wind from the east, which brings them to their shores. Those of Kamchatka receive ice from the same coast, which the eastern sea sends out in winter, two to three days in succession. One sees there, at certain times, flights of birds, which after a stay of a few months, return to the east, from where they arrive. The continent opposite to that of Asia in the north descends thus as far as the latitude of Kamchatka: this continent must be that of western America. M. Muller,* having given a summary of five or six journeys, which were attempted by the sea in the north to pass the northern point of Asia, finishes by saying that everything indicates the impossibility of such navigation; and he proves this by the following reasons: this navigation must be made in one summer; and, the interval from Archangel to the Ob, and from this river to Yenisei demands a good and entire season: the passage from Waeigats gave infinite trouble to the English and the Dutch. Leaving this glacial strait, one encounters islands that close the route. Then, the continent, which forms a cape between the rivers Piasida and Chatanga, projects beyond the seventy-sixth degree of latitude, and is bounded by a chain of islands, which make the passage difficult in navigation. If one wishes to move away from the coast and gain the high sea toward the pole, the mountains of almost-immobile ice found in Greenland and Spitsbergen—do they not indicate a continuity of ice as far as the pole? If one wishes to skirt the coast, *this navigation is less easy than it was a hundred years ago*: the water in the area has noticeably diminished there. One still sees, far from the coasts that are bathed by the glacial ocean, trees that it has thrown upon the land, which once served it as shoreline. These shores there are so shallow that one can use only very flat-bottomed boats there that, too feeble to resist the ice, are unable to provide for long navigation, nor are able to be loaded with the necessary provisions. Although the Russians have resources and the means that are not possessed by most other European countries to frequent these cold seas, one sees that the voyages attempted on the glacial sea have not yet opened up a route from Europe and from Asia to America; and it is only by leaving from Kamchatka or from another far eastern point on Asia that some coasts of western America were discovered.

Captain Bering left from the port of Awatscha in Kamchatka on 4 June 1741. After having traveled to the south-east and then back to the north-east, he saw, on the 18th of the following month, the continent of America at fifty-eight degrees and twenty-eighty minutes of latitude; two days later, he anchored near one island within a bay: seeing from

* *Histoire générale des voyages*, tome XVIII, 484.

there two capes, he called one to the east Saint-Élie, and the other to the west Saint-Hermogène. Then, he sent Chitrou, one of his officers, to reconnoiter and inspect the gulf that he had just entered. It was found to be intercut or sprinkled with islands. One of these possessed deserted huts: they were of well-assembled planks, even notched. It was conjectured that this island could have been inhabited by some people of the continent of America. M. Steller, sent to make observations upon these newly discovered lands, found a cave in which someone had placed provisions of smoked salmon, and left rope, furniture, and utensils. In the distance, he saw some Americans flee at his sight. Soon he saw a fire on a distant hill: the savages had doubtless retired there; a rock escarpment covered their retreat.

From the account of these facts, it is easy to judge that it can only be by departing from Kamchatka that the Russians can pursue commerce with China and Japan, and that for them it is as difficult, not to say impossible, as for other nations of Europe, to pass through the seas of the northeast, of which the greatest parts are entirely ice covered. Therefore, I do not fear to repeat that the sole possible passage is by the northwest, by Hudson Bay, and that this is the place where navigators must hold to, to find this passage that is so desirable and so evidently useful.

As I have already noted in the preceding pages of this volume, I received from M. le comte Schouvaloff, this great man of state that all Europe esteems and respects, I have received, say I, on the date of 27 October 1777, an excellent memoir composed by M. de Domascheneff, president of the Imperial Society of Petersburg, and to whom the Empress confided, in just title, the department of all that is connected to the sciences and the arts. This illustrious savant had at the same time sent me a hand-made copy of the map of the pilot Otcheredin, on which are represented the routes and the discoveries that he had made in 1770 and 1773, between the Kamchatka and the continent of America. M. de Domascheneff observes in his memoir that this map, of the pilot Otcheredin, is the most exact of all, and that the one that had been given in 1773 by the Academy of Petersburg needed to be modified in several places, notably in the position of islands and the supposed archipelago that had been represented there between the Aleute or Aleoute Islands and those of Anadir, otherwise called the Andrien Islands. The map of the pilot Otcheredin seems to show in effect that these two groups, the Aleute and the Andrien Islands, are separated by an open sea of more than a hundred leagues in extent. M. de Domascheneff assures us that the large general map of the empire of Russia, which has just been published in this year of 1777, represents exactly the coasts of all of the northern extremity of Asia inhabited by the Tschutschis: he said that this map had been prepared according to the most recent knowledge, acquired by the last expedition of Major Pawluzki against this people. "This coast," says M. de Domascheneff, "terminates the great chain of mountains that separates all of Siberia from southern Asia, and finishes in being divided between the chain that runs along the Kamchatka and those that occupy all the lands between the rivers that flow to the east of the Lena. The known islands between the coasts of Kamchatka and those of America are mountainous, as are those of Kamchatka and those of the continent of America. There is thus a well-marked continuity of the chains of mountains of the two continents, of which the interruptions, formerly perhaps less considerable, could have been enlarged by the wearing away of the rock, by the continual currents that come from the glacial sea toward the great sea of the south, and by the catastrophes of the globe."

But this submarine chain that joins the lands of Kamchatka with those of America is more to the south by seven or eight degrees than that of the Anadir or Andrien Islands which, from time immemorial, have served as passage for the Tschutschis to go to America.

M. de Domascheneff said that it is certain that this traverse from the point of Asia to the continent of America was made by rowing, and that these people went there to trade Russian ironware to the Americans; that the islands that are along this route are so near each other that one can sleep each night on solid ground, and that the continent of America where the Tschutschis traded is mountainous and covered with forests inhabited by foxes, martens, and sable, that yield furs of qualities and colors very different from those of Siberia. These northern islands situated between the two continents are scarcely known except by the Tschutschis; they form a chain between the easternmost point of Asia and the continent of America, below the sixty-fourth degree. This chain is separated by an open sea from a second chain to the south, of which we have spoken, situated below the fifty-sixth degree, between Kamchatka and America: these are the islands of this second chain that the Russians and the inhabitants of Kamchatka frequent to hunt sea otters and black foxes, of which the furs are very precious. People had knowledge of these islands, even of the easternmost ones in this last chain, before the year 1750: one of these islands carries the name of Commander Bering, another one fairly close by is called Medenoi island. Then, one finds the four islands of the Aleutes or Aleoutes, the first two situated a little above, and the last ones a little below the fifty-fifth degree. Then, one finds some at the fifty-sixth degree, the Atkhou and Amlaigh Islands, which are the first of the Renard chain of islands, which extends toward the northeast as far as the sixty-first degree of latitude: the name of these islands is taken from the prodigious number of foxes that one finds there. The two islands of Commander Bering and of Medenoi were uninhabited when they were discovered. But, there was found in the Aleute Islands, a little more advanced to the east, more than sixty families, of which the language was related neither to that of Kamchatka nor to any of those of east Asia, and is only a dialect of the language that is spoken in neighboring islands of America; this seems to indicate that they were peopled by Americans, and not by Asians.

The islands that were named, by the expedition of Bering, those of Saint-Julien, Sainte-Théodore, Saint Abraham, are the same as those that one calls today the Aleute Islands, and similarly, the islands of Chommaghin and Saint-Dolmat, indicated by this navigator, are part of those that one calls the Renard Islands.

"The great distance," says M. Domascheneff, "and the open and deep sea found between the Alcate islands and the Renard islands, together with the different rock foundations of the latter, can make one presume that these islands do not form a continuous marine chain, but that the former, with those of Medenoi and of Bering, make up a marine chain that comes from Kamchatka, and that the Renard Islands represent another one, coming from America; that one and the other of these chains in general become lost in the depths of the sea, and are promontories of the two continents. The continuation of the Renard Islands, of which some are of large extent, is intermixed with reefs and breakwaters, and continues without interruption as far as the continent of America. But, those that are closest to this continent are very little frequented by the barques of the Russian hunters, because they are well populated, and because it would be dangerous to make a sojourn there: there are several of these islands neighboring the terra firma of America that are not yet well known. A few ships have nevertheless penetrated as far as the island of Kadjak, which lies very close to the continent of America; one is assured of this as much by the reports of the islanders as by other reasons. One of these reasons is that, by contrast with the most western islands that only produce stunted and creeping shrubs, that the winds of the open sea prevent from growing taller, the island of Kadjak and the small neighboring islands produce groves of alder, which seems to indicate that they are less exposed, and that they are protected in the north and the east by a neighboring continent. Furthermore, one finds there freshwater otters, which are not at all seen on other islands, and also a small species of marmot, which appears to be the marmot of Canada; finally, one can see there evidence of bear and of wolf, and the inhabitants clothe themselves in reindeer skins, which come to them from the continent of America, which they are close neighbors of.

One sees, by the narrative of a voyage that pressed as far as the island of Kadjak, under the guidance of a certain Geottof, that the islanders give the name of *Atakthan* to the continent of America. They say that this great land is mountainous and all covered with forests; they place this great land to the north of their island, and call the mouth of a large river that is found there Alaghschak . . . On the other hand, one cannot doubt that Bering, as well as Tschirikov, effectively touched this great continent, since at Cape Élie, where his frigate was moored, it was seen that at the edge of the sea the terrain rises into mountains that were continuous and clothed in thick forests. The terrain there was of a nature very different to that of Kamchatka; a number of American plants were collected there by Steller."

M. de Domascheneff furthermore observed that all of the Renard islands, like those of the Aleutes and those of Bering, are mountainous, that their coasts are for the most part jagged with rocks, cut by precipices and surrounded by reefs for quite a distance; that the terrain rises from the coasts as far as the middle of these islands in mountains that are very steep, which form small chains along the length of each island. Furthermore, there were and there still are volcanoes on several of these islands, and those where the volcanoes are extinct possess hot springs. No metals can be found on these volcanic islands, but only chalcedonies and a few other colored stones of little value. There is no wood in these islands other than the trunks or branches of trees drifted in by the sea, and which do not arrive in large quantity; more is found on Bering island and on the Aleutes: it seems that these drifted trees come for the most part from southern shores, because one sees among them wood of camphor from Japan.

The inhabitants of these islands are fairly numerous, but as they lead a roving life, moving from one island to another, it is not possible to fix their number. It is generally observed that the longer the islands are, and the closer to America, the higher is the population. It also seems that all of the islanders of the Renard Islands are of the same nation, to which the inhabitants of the Aleutes and the Andrien islands could also be related, although they differ in certain customs. All of these people share a very great resemblance, by their customs, their way of life and of feeding themselves, with the Eskimos and the natives of Greenland. The name *Kanaghist* by which the islanders call themselves in their language, perhaps corrupted by the sailors, still closely resembles that of the *Karalit*, which the Eskimos and their brothers in Greenland call themselves. Among the inhabitants of all these islands, between Asia and America, the only tools they are found to have are stone axes, sharpened stones, and animal shoulder blades, sharpened to cut grass. They also have spears, which they throw by hand with the aid of a paddle, and of which the point is armed with a pointed stone that has been artistically shaped; today they have many iron implements stolen or taken from the Russians. They make canoes and types of dugout like those of the Eskimos; some are big enough to hold twenty people. The framework is of light wood, covered everywhere by the skins of seals and other marine animals.

It seems by all these facts that from times immemorial the Tschutchis, who inhabit the easternmost part of Asia, between fifty-five and seventy degrees, traded with the Americans, and that this commerce was all the easier for these people accustomed to the rigors of the cold, so that they could undertake this voyage, which is perhaps not a hundred leagues, by resting each day on one island and then another, in simple canoes, by rowing in the summer and maybe traveling across the ice in winter. America could thus have been peopled from Asia beneath the parallel; and everything seems to indicate that, although there are interruptions of sea between the land of these islands, they did not previously form anything other than one and the same continent, by which America was joined to Asia. This seems to indicate also that beyond the Anadir and Andrien Islands, that is to say, between the seventieth and seventy-fifth degrees, the two continents are absolutely united by terrain where there is no longer any sea, but which is perhaps entirely covered by ice. The reconnaissance of these shores beyond the seventieth parallel is an enterprise worthy of the attention of the great sovereign of Russia, one that would need to be confided to a navigator as courageous as M. Phipps. I am well persuaded that one would find these continents united; and if it is otherwise, and if there is open sea beyond the Andrien Islands, it seems to me certain that one would find there the extensions of the great glacier of the pole, at eighty-one or eighty-two degrees, as M. Phipps found them at the same height, between Spitsbergen and Greenland.

Notes on the Seventh Epoch

1. *The respect for certain mountains upon which they saved themselves from deluges; the horror of those other mountains that hurled out terrible fire, and so on.*

The revered mountains in the East are Mount Carmel, and a few places in the Caucasus: Mount Pirpangel to the north of Hindustan; Mount Pora in the province of Aracan; that of Chaq-pechan at the source of the Sangari river, where the Manchu Tartars live, from where the Chinese believe that Fo-hi came; Mount Altai to the east of the source of Selinga in Tartary; Mount Pecha to the northwest of China, and so on. Those which were held in horror were the volcanic mountains, among which one can cite Mount Ararath, of which the name signifies the mountain of unhappiness, because in effect this mountain was one of the largest volcanoes in Asia, something one can still recognize today by its form and by the materials that surround its summit, where one can see craters and other signs of its ancient eruptions.

2. *How could people so early have worked out the lunisolar period of six hundred years!*

The period of six hundred years, which Joseph said that the ancient patriarchs employed before the Deluge, is one of the most beautiful and most exact that has ever been invented. It is a fact that taking *the lunar month* of 29 *days* 12 *hours* 44 *minutes* 3 *seconds*, one finds that 219 *thousand,* 146 ½ *days* make 7 *thousand* 421 *lunar months*; *and the same number of* 219 *thousand,* 146 ½ *days* gives 600 *solar years, each of* 365 *days* 5 *hours* 51 *minutes* 36 *seconds*; from this results the solar month to within a second, as modern astronomers have determined it, and the solar year, more accurately than Hipparchus and Ptolemy determined it more than two thousand years after the Deluge. Joseph cited as authorities Manéthon, Bérose, and several other ancient writers, whose writings have long been lost . . . Whatever are the foundations upon which Joseph talked of this period, it must be that there was really, since time immemorial, such a period or great year, which had been forgotten for many centuries. Since the astronomers who came after this historian would have used it for preference to other, less exact hypotheses for the determination of the solar year and of the lunar month, if they had known it, or they would have been honored, if they would have imagined it.*

It is obvious, said the learned astronomer Dominique Cassini, that from the first age of the world, men had already made great progress in the science of the movement of stars. One could even propose that they even had far more knowledge of this than people had for a long time after the Deluge, if it is really true that the year that the ancient patriarchs used was of the grandeur of those that compose the great period of six hundred years, of which is made mention in the antiquity of the Jews, as written by Joseph. We do not find, in the monuments that remain to us from all the other nations, any vestige of this period of six hundred years, which is one of the most beautiful that has ever been invented.

M. Cassini refers, as one can see, to Joseph, and Joseph had as authorities Egyptian, Babylonian, Phoenician, and Greek historiographers: Manéthon, Berossus, Mochus, Hestieus, Jerome the Egyptian, Hesiod, Hecataeus, and others, whose writings could be preserved and seemingly were preserved in his time.

And this being proposed, and while one can oppose the witness of these writers, M. de Mairan said, with reason, that the incompetence of the judges or of the witnesses should not have place here. The facts themselves establish their authenticity: it suffices that a comparable period has been named; it suffices that it existed, so that one has the right to conclude that there would thus have existed centuries of observations, and a great number which preceded those: that the period of ignorance of this that followed is also very ancient; because one must regard as a time of ignorance all that time when people did not know of the correctness of this period, and when people neglected to deepen their knowledge of the elements and to make use of them, to amend the theory of celestial movements, and where one thinks to substitute less exact knowledge. Thus if Hipparchus, Melon, Pythagoras, Thales, and all the ancient astronomers of Greece were unaware of the period of six hundred years, one is justified in saying that it had been forgotten, not only among the Greeks, but also in Egypt, in Phoenicia, and in Chaldea, from where the Greeks had drawn all the great knowledge of astronomy.

3. *The Chinese, the Brahmins, any more than the Chaldeans, the Persians, the Egyptians, and the Greeks, received nothing from the first people who had so strongly advanced astronomy, and the beginnings of the new astronomy are due to the tenacious assiduity of the Chaldean observers, and then to the work of the Greeks.*

The Greek astronomers and philosophers had drawn most of their knowledge from Egypt and from India. The Greeks were thus people who were very new to astronomy by comparison with the Indians, the Chinese, and the Atlantean inhabitants of West Africa: Uranus and Atlas among these last peoples, Fo-hi in China, Mercury in Egypt, Zoroaster in Persia, and so on.

The Atlanteans, among whom Atlas reigned, seemed to have been the most ancient people of Africa, and much more ancient than the Egyptians. The *Theogony* of the Atlanteans, reported by Diodore of Sicily, was probably introduced into Egypt, into Ethiopia, and into Phoenicia at the time of this great exodus, which is spoken of in the *Timaeus*

* Letters of M. de Mairan to R. P. Parrenin, Paris, 1769, in-2, 108, 109.

of Plato, of an innumerable people who left the island of Atlantis and threw themselves upon a great part of Europe, of Asia, and of Africa.

In the west of Asia, in Europe, in Africa, everything is founded upon the knowledge of the Atlanteans, while the oriental people, the Chaldeans, Indians, and Chinese, were not educated until later, and were always people who had no relations with the Atlanteans, of which the exodus is more ancient than the first date of any of these lost peoples.

Atlas, the son of Uranus and brother of Saturn, lived, according to Manéthon and Dicaearchus, around 3,900 years before the Christian era.

While Diogenes Laërtius, Herodotus, Diodore of Sicily, and Pomponius Mela variously give from 48,600 to 23,000 years for the age of Uranus, that does not preclude reducing these years to their true measures of time that were used in different centuries among these peoples, thus resulting in the same one, namely 3,890 years before the Christian era.

The time of the Deluge, according to the Septantes, was 2,256 years after the creation.

Astronomy was cultivated in Egypt more than three thousand years before the Christian era; one can demonstrate this by what Ptolemy reports on the heliacal rising of Sirius; this rising of Sirius was very important among the Egyptians, because it announced the flooding of the Nile.

The Chaldeans seemed younger in astronomical study than the Egyptians.

The Egyptians knew the movement of the Sun more than three thousand years before Jesus Christ, and the Chaldeans more than 2,473 years.

There was among the Phrygians a temple dedicated to Hercules, which seems to have been founded 2,800 years before the Christian era, and one knows that in antiquity Hercules was the emblem of the Sun.

One can also date the astronomical knowledge of the ancient Persians at more than 3,200 years before Jesus Christ.

Astronomy in India is also as ancient; they allow four ages, and it is at the beginning of the fourth to which is tied their first astronomical epoch. This age had lasted, in 1742, for 4,863 years, which goes back to 3,102 years before Jesus Christ. This last age of the Indians is in reality composed of solar years, but the three others, of which the first is 1 million and 728 thousand years, the second 1 million 296 thousand, and the third of 864 thousand years, are evidently composed of years, or rather of revolutions of time very much shorter than solar years.

It is also shown by the astronomical epochs that the Chinese had cultivated astronomy more than three thousand years before Jesus Christ, and from the times of Fo-hi.

There is thus a kind of level between the Egyptian, Chaldean or Persian, Indian, Chinese, and Tartar peoples. None reach back more than others into antiquity, and this remarkable epoch of three thousand years of age for astronomy is about the same everywhere.*

4. *I could easily give several other examples, which all concur to demonstrate that man can modify the influences of climate where he lives.*

"Those who have long lived in Pennsylvania and in neighboring colonies have observed," says M. Hugues Williamson, "that their climate has changed considerably over forty or fifty years, and that the winters are nowhere near as cold . . .

* *Histoire de l'ancienne astronomie*, by M. Bailly.

"The temperature of the air in Pennsylvania is different from that of the countries of Europe that lie under the same parallel. To judge the warmth of a country, one must not only have regard to its latitude, but also to its situation and to the winds that have custom to prevail there. Because, these are not able to change without the climate changing also. The face of a country can be entirely metamorphosed by its farming; and one can convince oneself, in examining the cause of winds, that their course can, in like manner, take new directions . . .

"Since the establishment of our colonies," continues M. Williamson, "we have come not only to provide more heat to the terrain of the inhabited cantons, but to change in part the direction of the winds. The sailors, who are most interested in this affair, have told us that it used to take them four to five weeks to reach our coasts, while today they arrive here in half the time. One deduces therefore that the cold is less intense, the snow less abundant and less continuous than it has ever been since we have become established in this province . . .

"There are several other causes that can increase or diminish the heat in the air; but one cannot, however, cite a single example of change of climate that one cannot attribute to the clearance of the land where it took place. One can give me the example of someone who has arrived since 1,700 years ago in Italy or in a few countries of the East, as an exception to this general rule. It is said that Italy was better cultivated in the times of Augustus than it is today; and yet the climate there is much more temperate. It is true that the winters were more severe in Italy 1,700 years ago than they are today . . . ; but one can attribute the cause to the vast forests with which Germany, which is to the north of Rome, was covered in those times . . . Piercing north winds were raised in those wild wastelands, which spread like a torrent into Italy and caused excessive cold there, and the air was formerly so cold in these uncultivated regions that it must have ruined the equilibrium of the atmosphere in Italy, which is not the case in our days.

"One can thus reasonably conclude that in a few years from now, and when our descendants will have cleared the interior part of this land, they will hardly be subject any more to the ice or to the snow, and their winters will be much tempered."* These opinions of M. Williamson are very just, and I do not doubt that our posterity will see them confirmed by experience.

* *Journal de Physique*, by M. l'Abbé Rozier, June 1773.

Index